D0611334

METHODS in
MICROBIOLOGY

METHODS in

MICROBIOLOGY

Edited by
J. R. NORRIS
Agricultural Research Council,
Meat Research Institute,
Bristol, England.

Volume 9

 1976

ACADEMIC PRESS
London · New York · San Francisco
A Subsidiary of Harcourt Brace Jovanovich, Publishers

ACADEMIC PRESS INC. (LONDON) LTD
24–28 Oval Road
London NW1

U.S. Edition published by
ACADEMIC PRESS INC.
111 Fifth Avenue
New York, New York 10003

Library of Congress Catalog Card Number: 68–57745
ISBN: 0–12–521509–6

PRINTED IN GREAT BRITAIN BY
ADLARD AND SON LIMITED
DORKING, SURREY

LIST OF CONTRIBUTORS

D. B. DRUCKER, *Department of Bacteriology and Virology, University of Manchester, Manchester, England*

C-G. HEDÉN, *Karolinska Institutet, 5-104 01 Stockholm 60, Sweden*

T. ILLÉNI, *Karolinska Institutet, 5-104 01 Stockholm 60, Sweden*

D. KAY, *Sir William Dunn School of Pathology, University of Oxford, Oxford, England*

I. KÜHN, *Karolinska Institutet, 5-104 01 Stockholm 60, Sweden*

K. G. LICKFELD, *Laboratory for Electron Microscopy at the Institute of Medical Microbiology, Clinical Hospital, University of Essen (GHS), D-43 Essen, Federal Republic of Germany*

M. P. STARR, *Department of Bacteriology, University of California, Davis, California 95616, U.S.A.*

H. STOLP, *Lehrstuhl fur Mikrobiologie, Universitat Bayreuth, 8580 Bayreuth, Federal Republic of Germany*

R. R. WATSON, *Indiana University School of Medicine, Indianapolis, Indiana 46202, U.S.A.*

ACKNOWLEDGMENTS

For permission to reproduce, in whole or in part, certain figures and diagrams we are grateful to the following—

American Association for the Advancement of Sciences; American Chemical Society; American Society for Microbiology; S. Hirzel Verlag, Stuttgart; Preston Technical Abstracts Company, Illinois; Society of Chemical Industries, London; John Wiley and Sons, London; Karl Zeiss, Oberkochen, W. Germany.

Detailed acknowledgements are given in the legends to figures.

PREFACE

With the publication of Volume 8 of "Methods in Microbiology" the editors decided that they should not proceed immediately to gather material for further Volumes but that it was appropriate to wait a little and re-assess the developing field of microbial technology from time to time. During the intervening period, we have been approached by authors offering contributions and we have received a number of comments concerning areas which have not so far been adequately covered in the Series. These persuaded us that a further Volume should be produced but Professor Ribbons, because of developing interests elsewhere, decided that he could not devote the time to the production of a further Volume. I therefore decided to go ahead with a short Volume consisting of six contributions covering a number of fields selected partly on a basis of general interest and partly on the ground that they had not so far been covered.

The importance of electron microscopy in microbiology need hardly be emphasized and two Chapters concern this subject. The Chapter by Professor Lickfield is a general treatment of transmission electron microscopy with particular emphasis on the handling of micro-organisms and is suitable for the orientation of a newcomer to the field; Dr Kay's Chapter, on the other hand, deals with an application of electron microscopy of particular interest to the specialist.

Three Chapters concern rapid or novel methods for the characterization of micro-organisms which are finding growing application in the areas of microbial ecology and diagnosis. The remaining Chapter deals with the unusual parasitic Bdellevibrios, again a subject of general interest.

It is a pleasure for me to record the friendly co-operation that I have received from authors during the preparation of this Volume and my appreciation to Academic Press for their help throughout.

Several communicants have pointed out the need for a text detailing the methods of typing applied to various groups of bacteria and, after careful consideration of the merits of a text dealing with this important subject, I have embarked on a collaborative venture with Professor Tom Bergan, of the Microbiological Department of the Institute of Pharmacy in the University of Oslo, aimed at producing several further Volumes in the Series which will deal with specific groups of bacteria and the typing methods used for studying them. Manuscript material is at present coming in very well and I hope that a number of useful Volumes will appear in fairly rapid succession.

July, 1976
J. R. NORRIS

CONTENTS

CONTENTS

CONTENTS OF PUBLISHED VOLUMES

xi

Substrate Specificities of Aminopeptidases: A Specific Method for Microbial Differentiation

RONALD ROSS WATSON

Indiana University School of Medicine, Indianapolis, Indiana 46202, U.S.A.

I. DIFFERENTIATION BY ASSAY OF MICROBIAL AMINOPEPTIDASES

During the past few years the substrate specificities of aminopeptidases in various species of bacteria, parasites and fungi have been shown to be radically different. These differences have been used to distinguish closely related species and even strains of the same species (Huber *et al.*, 1970; Huber and Mullanax, 1969; Huber and Warren, 1975; Krawczyk and Huber, 1976; Mulczyk and Szewczyk, 1970; Peterson *et al.*, 1975; Uglem and Beck, 1972; Westley *et al.*, 1967; Lee *et al.*, 1975a; Perrine and Watson, 1975; Watson and Lee, 1974; Peterson and Hsu, 1974; Male, 1971; Muftic, 1967; Muftic, 1971). The aminopeptidase (arylamidase) procedure is based upon the enzymatic liberation of fluorescent β-naphthylamine (βNA) from a non-fluorescent L-amino-acid-β-naphthylamide. This allows quantitative and qualitative measurements as fluorescence is linear. Bacteria, fungi and parasites are differentiated on their ability to enzymatically hydrolyse a series of L-amino-acid-βNAs producing a specific, characteristic profile (see Fig. 1). Differentiation by substrate specificity of the aminopeptidases complements and sometimes supercedes morphological and cultural identification techniques. It does not suffer from data interpretational difficulties which are sometimes caused by interfering chemical constituents in electrophoretic, chromatographic, serological or phosphorescent techniques (Krawczyk and Huber, 1976). The biochemical basis of aminopeptidase identification avoids much of the time consuming growth in selective media which is a feature of identification by subjective morphological, cultural, physiological and pathogenic characteristics.

Recently aminopeptidases or aminopeptidase activity have been observed in multinucleated, pathogenic micro-organisms. Uglem and Beck (1972) were able to differentiate *Neolchinohynchus crassus* and *N. cristatus* which inhabit different parts of a fish intestine by the distinctive substrate specificities of their aminopeptidases. An interesting inverse relationship was found between the parasites' aminopeptidase activity, which is related to their pathogenicity, and that of the corresponding tissue they inhabit. Thus the strain of parasite with the lowest aminopeptidase activity inhabits the portion of the intestine which has the highest. It appears that they survive best where the combination of their own and host aminopeptidases produce an adequate supply of amino-acids (Uglem and Beck, 1972). Aminopeptidases have been observed in other parasites: *Paramecium caudatum* (Hunter, 1967), *Anisakis* species (Ruitenberg and Loendersloot, 1971) and *Ascaris suum* (Rhodes *et al.*, 1966).

Aminopeptidase activity has been routinely found in smaller, nucleated organisms. Many fungi contain aminopeptidases as shown in Table I. Aminopeptidases may well be found in all fungi, certainly they are found in fungi pathogenic for man (Lee *et al.*, 1975a), non-pathogenic fungi (Male, 1971), in the mycelium form (Lee *et al.*, 1975a), in the yeast form (Lee *et al.*, 1975a; Male, 1971; Reiss, 1971) and in basidiomycetes (Blaich, 1973a, b). In *Histoplasma capsulatum* Watson and Lee (1974) found that aminopeptidases were exclusively intracellular while Labbe, Rebeyrotte and Turpin (1974) showed the presence of extracellular aminopeptidases in *Aspergillus oryzae*. Lee, Reca, Watson, and Campbell (1975) differentiated 8 human pathogenic yeasts based upon their aminopeptidase substrate specificities. Earlier Huber and Mullanax (1969) showed the presence of aminopeptidase activity in fungi pathogenic for plants by their ability to hydrolyse various L-amino-acid-βNAs. These fungi, *Fusarium oxysporum*, *F. solani* and *F. roseum* were differentiated on the basis of the substrate specificities of their aminopeptidases. *F. solani* was the only one to hydrolyse significantly L-prolyl-βNA and L-hydroxyprolyl-βNA. The other two fungi differed less, with quantitative differences in hydrolysis

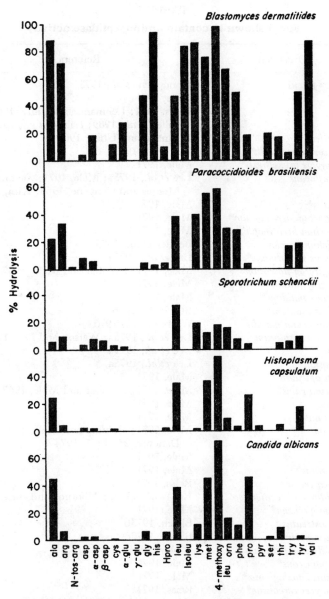

FIG. 1. Aminopeptidase profiles of five yeasts, *Histoplasma capsulatum*, *Sporotri-chum schenckii*, *Candida albicans*, *Blastomyces dermatitidis*, and *Paracoccidioides brasiliensis*. These profiles were obtained as described above for the whole cell technique.

TABLE I

Fungi shown to contain aminopeptidase activity

Organism	References
Aspergillus flavus	Konoplich *et al.*, 1973
*A. niger**	Male, 1971
A. oryzae	Reiss, 1971; Lehmann and Uhlig, 1969; Jolles *et al.*, 1969; Labbe *et al.*, 1974
A. parasiticus	Lehmann and Uhlig, 1969
*Blastomyces dermatitidis**	Lee *et al.*, 1975a
*Brettanomyces bruxellinsis**	Male, 1971
Candida albicans	Lee *et al.*, 1975a; Male, 1971; Zaikina, 1969; Montes and Constine, 1968; Kim, 1962
*C. tropicalis**	Male, 1971
*Cephalosporium acremonium**	Male, 1971
*Chrysosporium keratinophilus**	Male, 1971
Colletotrichum species	Bender *et al.*, 1975
*Cryptococcus neoformans**	Lee *et al.*, 1975a
*Epidermophyton floccosum**	Male, 1971; Holubar and Male, 1967
*Fusarium oxysporum**	Male, 1971
*Geotrichum candidum**	Male, 1971
Hapalopilus nidulans	Blaich, 1973a, b
Helminthosporium maydis	Warren *et al.*, 1976
Histoplasma capsulatum	Lee *et al.*, 1975a; Watson and Lee, 1974
H. dubosii	Lee *et al.*, 1975a
*H. farcinosum**	Lee *et al.*, 1975a
*Kloeckera apiculata**	Male, 1971
*Microsporum canis**	Male, 1971; Holubar and Male, 1967
M. cookei	Male, 1971
*M. distortum**	Male, 1971
M. gypseum	Male, 1971; Holubar and Male, 1967; Danew *et al.*, 1971, 1974
*M. nanum**	Male, 1971
*Mucor pusillus**	Male, 1971
*Neurospora crassa**	Reiss, 1971
Paracoccidioides brasiliensis	Lee *et al.*, 1975a; Albornoz and Aasen, 1971
*Penicillium glaucum**	Male, 1971
Pleurotus ostreatus	Blaich, 1973b
Podospora anserina	Blaich, 1973c
Polysphondylium pallidum	O'Day, 1974
*Rhizopus nigricans**	Male, 1971
*Rhodotorula mucilaginosa**	Male, 1971
*Saccharomyces cerevisiae**	Reiss, 1971
S. lactis	Desmazeaud and Devoyod, 1974
Schizosaccharomyces pombe	Rock and Johnson, 1970
*Scopulariopsis brevicaulis**	Male, 1971
*Stremphylium sarcinaeforme**	Male, 1971
Thamnidium anomalum	Stout and Shaw, 1973

TABLE I—*cont.*

Organism	References
T. elegans	Stout and Shaw, 1973
*Torulopsis famata**	Male, 1971
*Trichophyton guinekeanum**	Holubar and Male, 1967
*T. granulosum**	Male, 1971
*T. interdigitale**	Male, 1971
*T. mentagrophytes**	Holubar and Male, 1967
*T. persicolor**	Male, 1971
T. rubrum	Male, 1971; Holubar and Male, 1967; Danew *et al.*, 1971
*T. schonlenii**	Holubar and Male, 1967
T. terrestre	Male, 1971
T. tonsurans	Male, 1971
*T. verrucosum**	Male, 1971; Holubar and Male, 1967
*T. violaceum**	Holubar and Male, 1967
*Trichosporon cutaneum**	Male, 1971
*Trichothecium roseum**	Male, 1971
*Trigonopsis variabilis**	Male, 1971
Tritirachium album	Hennrich *et al.*, 1973
*Verticillium cinnabarium**	Male, 1971

* Organism whose aminopeptidase activity was measured without isolation or characterization of their aminopeptidase(s). Organisms not starred are those from which aminopeptidases have been isolated.

of L-γ glutamyl-βNA, L-alanyl-βNA, L-methionyl-βNA, L-phenylalanyl-βNA and L-threonyl-βNA making their differentiation rapid and unequivocal. The substrate specificities of the aminopeptidases in some fungi are clearly related to their amino-acid growth requirement (Lee *et al.*, 1975b, c). The distinctive aminopeptidase profiles of some yeasts *Candida albicans, Histoplasma capsulatum, Blastomyces dermatitidis, Paracoccidioides brasiliensis* and *Cryptococcus neoformans* have been successfully used to determine the N-terminal L-amino acids most rapidly liberated by the aminopeptidases of these organisms (Lee *et al.*, 1975b, c). The most rapidly liberated amino-acids are required for growth or differentiation.

As shown in the Table II aminopeptidases have been found often in bacteria. Their distinctive profiles have been used to distinguish closely related bacteria such as the *Neisseria* (Perrine and Watson, 1975), *Leptospira* (Burton *et al.*, 1970) and some *Enterobacteriaceae* (Petersen *et al.*, 1975; Petersen and Hsu, 1974). Some of this distinctive aminopeptidase activity may be related to pathogenicity or survival of the pathogen in the special environment of the host. Hence the aminopeptidase specificities may be quite distinctive in pathogens (Huber *et al.*, 1970; Perrine and Watson,

TABLE II
Bacteria shown to contain aminopeptidase activity

Organism	Reference
Aeromonas proteolytica	Litchfield and Prescott, 1970; Prescott *et al.*, 1971
Arthrobacter species	Ryden, 1971; Bjare *et al.*, 1970
Bacillus anthracis	Muftic and Schroder, 1966
B. brevis*	Westley *et al.*, 1967
B. cereus	Westley *et al.*, 1967; Muftic and Schroder, 1966
B. circulans*	Westley *et al.*, 1967
B. globigii*	Westley *et al.*, 1967
B. laterosporus*	Westley *et al.*, 1967
B. lentus	Westley *et al.*, 1967; Muftic and Schroder, 1966
B. licheniformis	Westley *et al.*, 1967; Muftic and Schroder, 1966; Davies, 1969
B. megaterium	Westley *et al.*, 1967; Muftic and Schroder, 1966; Aubert and Miller, 1965
B. mesentericus	Muftic and Schroder, 1966
B. polymyxa*	Westley *et al.*, 1967
B. pumilus	Westley *et al.*, 1967; Muftic and Schroder, 1966
B. rotans*	Westley *et al.*, 1967
B. sphaericus*	Westley *et al.*, 1967
B. subtilis	Westley *et al.*, 1967; Muftic and Schroder, 1966; Takeda and Webster, 1969; Ray and Wagner, 1972; Matsumura *et al.*, 1971; Roncari and Zuber, 1969; Moser *et al.*, 1970; Roncari *et al.*, 1972; Stoll *et al.*, 1972; Zuber and Roncari, 1967
Bacteroides amylophilus	Blackburn, 1968
Bombina variegata	Molzer and Michi, 1967
Clostridium histolyticum	Kessler and Yaron, 1973
Enterobacter aerogenes	Petersen *et al.*, 1975; Petersen and Hsu, 1974
Erwinia amylovora	Krawczyk and Huber, 1976
Escherichia coli	Westley *et al.*, 1967; Petersen and Hsu, 1974; Takeda and Webster, 1969; Tsai and Matheson, 1965; Vogt, 1970; Brown and Krall, 1971; Simmonds, 1971; Miller and Prescott, 1972
Klebsiella cloacae	Kwiatkowska *et al.*, 1974
Lactobacillus acidophilus*	Oya *et al.*, 1971
L. casei*	Oya *et al.*, 1971
Leptospira autumnalis (and 13 other species)	Burton *et al.*, 1970
mycobacteria species	Muftic, 1967, 1971

TABLE II—*cont.*

Organism	Reference
Mycoplasma laidlawii	Choules and Gray, 1971
Neisseria catarrhalis	Behal and Folds, 1967; Cox and Behal, 1969; 1970
*N. flavescens**	Perrine and Watson, 1975
*N. gonorrhoeae**	Perrine and Watson, 1975
*N. meningitidis**	Perrine and Watson, 1975
*N. sicca**	Perrine and Watson, 1975
Picea abies	Lundvist, 1974
Pseudomonas aeruginosa	Riley and Behal, 1971
P. tabaci	Krawczyk and Huber, 1976
Sarcina lutea	Behal and Cater, 1971
Salmonella enteritidis	Petersen *et al.*, 1975; Petersen and Hsu, 1974
S. typhimurium	MacHugh and Miller, 1974
Staphylococcus aureus	Zaikina, 1969
*Streptococcus faecalis**	Oya *et al.*, 1971
S. mites	Oya *et al.*, 1971; Linder, 1974; Linder *et al.*, 1974
S. pneumoniae	Johnson, 1973
*S. salivarius**	Oya *et al.*, 1971
S. thermophilus	Rabier and Desmazeaud, 1973; Myrin and Hofsten, 1974; Hengartner *et al.*, 1973; Balerna and Zuber, 1974
Streptomyces fradiae	Morihara *et al.*, 1967
S. griseus	Vosbeck *et al.*, 1973
Xanthomonas campestris	Krawczyk and Huber, 1976

* Organisms whose aminopeptidase activity was measured without isolation or characterization of their aminopeptidase(s). Organisms not starred are those from which aminopeptidases have been isolated.

1975; Burton *et al.*, 1970). For example, Burton, Blenden and Goldberg (1970) found that the pathogenic strains of *Leptospira* had more aminopeptidase activity which hydrolysed L-leucyl-βNA and less which hydrolysed L-alanyl-βNA than the non-pathogenic strains.

Finally Bouillant and Daniel (1968) report that even viruses can be detected or differentiated by the aminopeptidase profile of their host mammalian cells. Significant differences in the activity of leucine aminopeptidase were observed between KB cells infected with three strains of *Myxovirus parainfluenzae* and uninfected cells. The increased aminopeptidase activity seems to be related to the virus infection. It was found to be reliable in detecting *M. parainfluenzae* infection in this cell line.

II. AMINOPEPTIDASE SUBSTRATE SPECIFICITIES
MEASURED IN WHOLE CELLS

The *in vivo* assay procedure given below as the model technique was recently reported by Lee, Reca, Watson and Campbell (1975a) to differentiate human pathogenic yeasts. Only two modifications are necessary to apply this technique to other groups of organisms for identification of unknown species within the group. They are (a) selection of a growth media suitable to sustain growth of every member of the group and (b) determination of the inoculum and incubation time. The incubation time and the size of inoculum combined should be sufficient to produce significant (>20%) hydrolysis of one or more substrates with each organism. It is preferable that these conditions do not result in near total hydrolysis of most substrates cleaved by any organism's aminopeptidases, as this reduces the reproducibility of the assay.

A. Preparation of Yeast Cell Suspension for Aminopeptidase Assay

Yeast cell suspensions are prepared by first transferring them to Brain Heart Infusion agar slants (BHI) supplemented with 0·1% cysteine and 1% glucose and incubating them at 37°C for 48 h. The total growth is washed from 2 slants and transferred to 50 ml BHI broth and shaken for an additional 48 h. These log phase yeasts are harvested by centrifugation at $5000 \times g$ and washed twice with sterile minimal salt solution (MSS) containing glucose. MSS is prepared as follows—

Minimal salt solutions

K_2HPO_4	7·0 g
KH_2PO_4	3·0 g
Sodium citrate	0·5 mg
$MgSO_4.7H_2O$	0·2 g
$(NH_4)_2SO_4$	2·0 g

to 1000 ml water and sterilized by autoclaving. Before use a solution of glucose in distilled water (250 mg/litre) previously sterilized is added to MSS to produce a final concentration of glucose in the solution of 25 mg/litre.

Standardized inoculation suspension for the aminopeptidase assay is prepared by resuspending the washed yeast in sterile MSS plus 2·5% glucose and 0·02% Gel Gard (fine grind, Dow Chemical Co., Midland, Michigan, USA) in round cuvettes, 12×75 mm adjusting the suspension to 25% transmission at 550 nm with a spectrophotometer. Gel Gard maintains the yeasts in suspension thus minimizing pipetting errors. The 2·5% glucose is used to partially minimize the quenching effect of chlorine in the Tris-HCl buffer used to prepare the substrates (Huber and Mullanax, 1969).

B. Preparation of Amino-acid-β-naphthylamide Substrates

The 26 amino-acid-beta-naphthylamide derivatives are listed below.

L-Alanyl-β-naphthylamide
L-Arginyl-β-naphthylamide
N-Tosyl-L-arginyl-β-napthylamide
L-Asparagyl-β-naphthylamide (incompletely dissolved)
L-alpha-aspartyl-β-naphthylamide
L-beta-aspartyl-β-naphthylamide (incompletely dissolved)
L-Cystine-D, L-β-naphthylamide (incompletely dissolved)
L-alpha-glutamyl-β-naphthylamide
L-gamma-glutamyl-β-naphthylamide (sterilized by filtration)
Glycyl-β-naphthylamide
L-Histidyl-β-naphthylamide
L-Hydroxyl-prolyl-β-naphthylamide
L-Leucyl-β-naphthylamide
L-Isoleucyl-β-naphthylamide
L-Lysyl-β-naphthylamide
L-Methionyl-β-naphthylamide
4-Methoxy-leucyl-β-naphthylamide
L-Ornithyl-β-naphthylamide
L-Phenylalanyl-β-naphthylamide
L-Prolyl-β-naphthylamide
L-Pyrolidonyl-β-naphthylamide
L-Seryl-β-naphthylamide
L-Threonyl-β-naphthylamide
L-Tryptophyl-β-naphthylamide
L-Tyrosyl-β-naphthylamide
L-Valyl-β-naphthylamide

Most of the above compounds can be obtained from Mann Research Laboratories, Division of Becton, Dickinson & Co., New York, New York. They are also available from Sigma Chemical Company, St. Louis, Missouri, U.S.A., or Nutritional Biochemical Corporation, Cleveland, Ohio, U.S.A.

These substrates are used at a concentration of 1×10^{-4}M and are prepared precisely as described by Huber and Mulanax (1969) and Krawczyk and Huber (1976). Each L-amino-acid-β-naphthylamide (amino-acid-βNA) is placed in 0·1 M Tris-HCl buffer, pH 8·0 at a concentration of 1×10^{-4} M and autoclaved for 10 min at 15 psi. An exception is L-gamma-glutamyl-βNA which is filter sterilized as this substrate hydrolyses during autoclaving. The sterile substrates are pipetted aseptically in 2·0 ml quantities into sterile 12×75 mm clear polystyrene tubes (Falcon No. 2003). If the tubes are capped they can be stored at 4°C for months before use.

C. Profile Determination with Whole Yeast Cells

To 2 ml of each substrate warmed to 37°C in a water bath, 0·2 ml standardized yeast inoculum is added. The tubes are recapped and incubated for 1 h at 37°C. (The use of capped tubes is suggested with potential human pathogens. Uncapped glass or quartz cuvettes can be used with non-pathogens. The use of polystyrene tubes increased the error in reading insignificantly.) Beta-naphthylamine and Tris-HCl buffer are also inoculated as controls. A set of the amino-acid-βNA derivatives without inoculum are also incubated as controls for any internal hydrolysis which is not seen unless the substrates are contaminated. Immediately after incubation, enzymatic activity is measured by the fluorescence of the released beta-naphthylamine on a Turner Model IIK-000 fluorometer with a Corning 7-60 as primary filter and the combination of a Wratten 47B filter, a 1% filter and (if needed to reduce fluorescent intensity) a neutral density filter as secondary filters. The fluorometer is calibrated prior to the reading of the tests, using Tris-HCl buffer as zero per cent control and the fluorescence of 1×10^{-4}M beta-naphthylamine as 100% hydrolysis control.

III. AMINOPEPTIDASE SUBSTRATE SPECIFICITIES MEASURED WITH SOLUBLE ENZYMES

The above *in vivo* technique using viable whole organisms has been modified successfully by Petersen, Sell and Hsu (1975). This modification has the advantage that the amino-acid-β-naphthylamides do not have to enter the cell to be hydrolysed by soluble, cell-free, enzymes. The use of cell-free aminopeptidases in place of an *in vivo* cellular source of the enzymes, although somewhat more cumbersome, has the advantage of giving a more complete profile. Some amino-acid-βNAs are not transported into the cell for hydrolysis. Both methods produce different, although completely reproducible profiles.

A. Preparation of Cell-free Aminopeptidases from
Enterobacteriaceae

Each *Enterobacteriaceae* culture is actively grown in Trypticase Soy Broth (TSB) at 37°C with shaking. This culture is diluted 1 : 35 (v/v) in fresh TSB. Incubation of the diluted culture in a shaker water bath at 37°C is continued until the optical density reaches 1·5 at 600 nm. The culture is cooled rapidly and kept in an ice bath to 4°C for the rest of the enzyme isolation. The cells are washed three times by centrifugation at $1500 \times g$ for 10 min with cold saline (0·9% NaCl). The bacteria are resuspended in saline and adjusted to the concentration producing an optical density of 1·5 at 500 nm. Sonication of this suspension in an ice bath for 30 second intervals with 3 min for cooling is performed until more than 90% of the cells are disrupted. Rupture of the bacterial cell wall is monitored by phase-contrast microscopy.

Modifications of these growth conditions are sometimes necessary depending upon the group of organisms being differentiated. Only complex media such as TSB and BHI have been used for growing bacteria or fungi for identification through their aminopeptidase profile. These media contain polypeptides as well as small peptides such as di- and tri-peptides which may be inducers of aminopeptidases. Synthetic media whose nitrogen source contain no polypeptides and only those amino-acids necessary for growth or differentiation such as is described for yeasts (Lee *et al.*, 1975b, c) have not been tested for their aminopeptidase profiles. These media may be even more distinctive for differentiation based upon the constitutive aminopeptidases. Quantitatively 10 times more amino-peptidase activity was induced in *Aeromonas proteolytica* when the nitrogen source in the growth media was enzymatically hydrolysed casein rather than a simple nitrogen source (Litchfield and Prescott, 1970). Litchfield and Prescott (1970) also found that the production of aminopeptidases was greatly enhanced by the addition of small quantities of peptide mixtures to the media containing a simple nitrogen compound as the primary nutrient source.

B. Profile Determination with Cell-free Aminopeptidases

Each L-amino-acid-βNA is prepared as described above. Each substrate (2 ml) is inoculated with 0·1 ml of the *Enterobacteriaceae* enzyme preparation in glass cuvettes and mixed. Smaller glass disposable tubes can be used with 0·2 ml of substrate and 10 μl of enzymes. The reaction mixtures are incubated for 1 h at 37°C in a water bath and hydrolysis stopped by heating to boiling for 5 min. Prior to assay of naphthalamine production, each sample is cooled to 4°C to restore maximum fluorescence. A blank of each substrate is inoculated with 0·1 ml of saline when the enzyme preparation is inoculated and otherwise treated identically. Its fluorescence is substracted from that produced by the enzyme inoculated preparation.

Both methods have been applied successfully to differentiating closely related species of micro-organisms. A recent study by Krawczyk and Huber (1976) showed that temperature of incubation, presence of metal ions, cofactors or chelators had little effect on the profiles obtained. Prior growth media, inoculum density and incubation time could have a great influence on aminopeptidase activity in the plant pathogens they studied. However, if these conditions are strictly controlled then the profiles obtained are highly reproducible and diagnostic. In an increasing variety of organisms including subspecies of kingsnakes (Dessaur and Pough, 1975), species of wheat (Reeve and Huber, 1975), parasites (Uglem and Beck, 1972), fungi (Lee *et al.*, 1975a; Bender *et al.*, 1975; Warren *et al.*, 1976) bacteria (Watson and Lee, 1974), different races of fungal spores (Huber *et al.*, 1975) and virus infected and non-infected mammalian cells (Bouillant and Daniel, 1968) aminopeptidase activity and substrate specificities have been used to distinguish members of a group of similar organisms.

ACKNOWLEDGEMENTS

The author wishes to recognize the advice and aid provided by Dr D. M. Huber of Purdue University, Dr R. D. Watson of the University of Idaho and E. Petersen of the Food and Drug Administration.

REFERENCES

Albornoz, M. B., and Aasen, I. C. (1971). *Sabouraudia*, **9**, 139–143.
Aubert, J., and Miller, J. (1965) *C. R. H. Acad. Sci.*, **261**, 4274–4277.
Balerna, M., and Zuber, H. (1974). *Int. J. Pept. Protein Res.*, **6**, 499–514.
Behal, F. J., and Carter, R. T. (1971). *Can. J. Microbiol.*, **17**, 39–45.
Behal, F. J., and Folds, J. D. (1967). *Arch. Biochem. Biophys.*, **121**, 364–371.
Bender, J. M., Stevenson, W. R., Keller, R. R. and Huber, D. M. in press. *Indiana Acad. Sci.*, **84**.
Bjare, U., Hofsten, B. N., and Ryden, A. C. (1970). *Biochin Biophys. Acta.* **220**, 134–136.
Blackburn, T. H. (1968). *J. Gen. Microbiol.*, **53**, 37–51.
Blaich, R. (1973a). *Arch. Mikrobiol.*, **86**, 118–128.
Blaich, R. (1973b). *Arch. Mikrobiol.*, **88**, 111–118.
Blaich, R. (1973c). *Arch. Mikrobiol.*, **94**, 201–220.
Bouillant, A., and Daniel, P. (1968). *Can. J. Microbiol.*, **14**, 971–974.
Brown, J. L., and Krall, J. F. (1971). *B.B.R.C.*, **42**, 390–397.
Burton, G., Blenden, D. C., and Goldberg, H. S. (1970). *Appl. Microbiol.*, **19**, 586–588.
Choules, G. L., and Gray, W. R. (1971) *B.B.R.C.*, **45**, 849–855.
Cox, S. T., and Behal, F. J. (1969). *Can. J. Microbiol.*, **15**, 1293–1300.
Cox, S. T., and Behal, F. J. (1970). *Proc. Soc. Exp. Biol. Med.*, **133**, 1247–1249.
Danew, P., Friedrich, E., Iwig, M., and Hanson, H. (1974) *Mykosen*, **17**, 179–189.
Danew, V. P., Friedrich, E., and Mannsfeldt, H. G. (1971). *Derm. Mschr.*, **157**, 232–238.
Davies, J. W. (1969). *Biochim. Biophys. Acta*, **174**, 686–695.
Desmazeaud, M. J., and Devoyod, J. J. (1974). *Ann. Biol. Anim. Biochim. Biophys.*, **14**, 327–341.
Dessauer, H. D., and Pough, F. H. (1975). *Comp. Biochem. Physiol.*, **50B**, 9–12.

Hengartner, H., Stoll, E., and Zuber, H. (1973). *Experientia*, **29**, 941–942.
Hennrich, N., Klockow, M., Orth, H. D., Femkert, U., Cichocki, P., Jany, K. (1973). *Hoppe Seylers Z. Physiol. Chem.*, **354**, 1529–1540.
Holubar, V. K., and Male, O. (1967). *Acta Histochem.*, **27**, 303–308.
Huber, D. M., and Mullanax, M. W. (1969). *Phytopathology*, **59**, 1032–1033.
Huber, D. M., and Warren, H. L. (1975). *Phytopathology*, **65**,
Huber, D. M., Guthrie, J. W., and Burnvik, O. (1970). *Phytopathology*, **60**, 1534 (Abstract).
Huber, D. M., Finney, R. E., and Shaner, G. E. (1975). *Proc. Am. Phytopath. Soc.*, **2**, 86.
Hunter, M. W. (1967). *Can. J. Microbiol.*, **13**,
Johnson, M. K. (1973). *Antonie Van Leeuwenhoek*, **39**, 599–608.
Jolles, J., Jolles, P., Uhlig, H., and Lehmann, K. (1969). *Hoppe-Seyler's Z. Physiol. Chem.*, **350**, 139–142.
Kessler, E., and Yaron, A. (1973). *B.B.R.C.*, **50**, 405–412.
Kim, Y. P. (1962). *J. Invest. Derm.*, **38**, 115.
Konoplich, L. A., Tsyperovich, A. S., and Kolodzeiskaia, M. V. (1973). *Ukrc. Ukr. Biokhim. Zh.*, **45**, 161–165.
Krawczyk, K., and Huber, D. M. (1976). Indiana Academy of Science. In press.
Kwiatkowska, J., Torain, B., and Glenner, G. G. (1974). *J. Biol. Chem.*, **249**, 7729–7736.
Labbe, J. P., Rebeyrotte, P., and Turpin, M. (1974). *C. R. Acad. Sc. Paris*, **278D**, 2699–2702.
Lee, K. L., Reca, M. E., Watson, R. R., and Campbell, C. C. (1975a). *Sabouraudia*, **13**, 132–141.
Lee, K. L., Reca, M. E., and Campbell, C. C. (1975b). *Sabouraudia*, **13**, 142–147.
Lee, K. L., Buckley, H. R., and Campbell, C. C. (1975c). *Sabouraudia*, **13**, 148–153.
Lehmann, V. K., and Uhlig, H. (1969). *Hoppe-Seyler's Z. Physiol. Chem.*, **350**, 99–104.
Linder, L. (1974). *Acta Pathol. Microbiol. Scand.*, **82B**, 593–601.
Linder, L., Lindquist, L., Soder, P. O., and Holme, T. (1974). *Acta Pathol. Microbiol. Scand.*, **82B**, 602–607.
Litchfield, C. D., and Prescott, J. M. (1970). *Can. J. Bacteriol.*, **16**, 23–27.
Lundvist, K. (1974). *Hereditas*, **76**, 91–95.
MacHugh, G. L., and Miller, C. G. (1974). *J. Bacteriol.*, **120**, 364–371.
Male, O. (1971). *Zbl. Bakt. I. Abt. Orig.*, **217**, 111–127.
Matsumura, Y., Minamiura, N., Fukymoto, J., and Yamamoto, T. (1971). *Agr. Biol. Chem.*, **35**, 975–982.
Miller, J. M., and Prescott, J. M. (1972). *Int. J. Pept. Protein Res.*, **4**, 415–419.
Molzer, H., and Michl, H. (1967). *Toxicon*, **5**, 105–109.
Montes, L. F., and Constantine, U. S. (1968). *J. Invest. Derm.*, **51**, 1–3.
Morihara, K., Oka, T., and Tsuzuki, H. (1967). *Biochim. Biophys. Acta*, **139**, 382–397.
Moser, P., Roncari, G., and Zuber, H. (1970). *Int. J. Protein Res.*, **2**, 191–207.
Muftic, M. (1967). *Folia Microbiol.*, **12**, 500–507.
Muftic, M. (1971). *Jap. J. Tuberc. Chest Dis.*, **17**, 31–34.
Muftic, M., and Schroder, E. (1966). *Path. Microbiol.*, **29**, 252–256.
Mulczyk, M., and Szewczyk, A. (1970). *J. Gen. Microbiol.*, **61**, 9–13.
Myrin, P., and Hofsten, B. V. (1974). *B.B.A.*, **350**, 13–25.
O'Day, D. H. (1974). *Dev. Biol.*, **36**, 400–410.

Oya, H., Nagatsu, T., Kobayashi, Y., and Takei, M. (1971). *Arch. Oral. Biol.*, **16**, 675–680.

Perrine, S., and Watson, R. R. (1975). Abstracts of the Annual Meeting of the American Society of Microbiology, p. 39.

Petersen, E. H., and Hsu, E. J. (1974). Abstracts of the Annual Meeting of the American Society of Microbiology, p. 24.

Petersen, E. H., Sell, S. L., and Hsu, E. J. (1975). Abstracts of the Annual Meeting of the American Society of Microbiology, p. 202.

Prescott, J. M., Wilkes, S. H., Wagner, F. W., and Wilson, K. J. (1971). *J. Biol. Chem.*, **246**, 1756–1764.

Rabier, D., and Desmazeaud, M. J. (1973). *Biochemie*, **55**, 389–404.

Ray, L. E., and Wagner, F. W. (1972). *Can. J. Microbiol.*, **18**, 853–859.

Reeve, R., and Huber, D. M. (1975). *Proc. Am. Phytopath. Soc.*, **2**, 88.

Reiss, V. J. (1971). *Acta Histochem.*, **39**, 277–285.

Rhodes, M. B., Marsh, C. L., and Ferguson, D. L. (1966). *Exp. Parasit.*, **19**, 42–51.

Riley, P. S., and Behal, F. J. (1971). *J. Bact.*, **108**, 809–816.

Rock, G. D., and Johnson, B. F. (1970). *Can. J. Microbiol.*, **16**, 187–191.

Roncari, G., and Zuber, H. (1969). *Int. J. Protein Res.*, **1**, 45–61.

Roncari, G., Zuber, H., and Wyttenbach, A. (1972). *Int. J. Protein Res.*, **4**, 267–271.

Ruitenberg, E. J., and Loendersloot, H. J. (1971). *J. Parasitol.*, **57**, 1149–1150.

Ryden, A. C. (1971). *Acta Chem. Scand.*, **25**, 847–858.

Simmonds, S. (1971). *Ciba Found. Symp.* 43–57.

Stoll, E., Hermodson, M. A., Ericsson, L. H., and Zuber, H. (1972). *Biochemistry*, **11**, 4731–4735.

Stout, D. L., and Shaw, C. R. (1973). *Mycologia*, **65**, 803–808.

Takeda, M., and Webster, R. E. (1969). *Proc. Nat. Acad. Sci.*, **60**, 1487–1494.

Tsai, C. S., and Matheson, A. T. (1965). *Can. J. Biochem.*, **43**, 1643–1652.

Uglem, G. L., and Beck, S. M. (1972). *J. Parasit.*, **58**, 911–920.

Vogy, V. M. (1970). *J. Biol. Chem.*, **245**, 4760–4769.

Vosbeck, K. D., Chow, K. F., and Awad, W. J., Jr. (1973). *J. Biol. Chem.*, **248**, 6029–6034.

Warren, H. L., Jones, A. Jr., and Huber, D. M. (1976). *Mycologia*, in press.

Watson, R. R., and Lee, K. L. (1974). Abstracts of the Annual Meeting of the American Society of Microbiology, p. 144.

Westley, J. W., Anderson, P. J., Close, V. A., Halpern, B., and Lederberg, E. M. (1967). *Appl. Microbiol.*, **15**, 822–825.

Zaikina, N. A. (1969). *Antonie van Leeuwenhoek*, **35**, 125–126.

Zuber, H., and Roncari, G. (1967). *Angew. Chem.*, **79**, 906–907.

CHAPTER II

Mechanized Identification of Micro-organisms*

CARL-GÖRAN HEDÉN, TIBOR ILLÉNI AND INGER KÜHN

Karolinska Institutte, S–104 01 Stockholm 60, Sweden

Traditional methods for the identification of micro-organisms are difficult to quantify. This is true also for the biochemical tests which often require so much manual labour that it is difficult to make full use of numerical taxonomy.

The "Autoline"-approach is based on optical scanning of 1 by 50 cm agar strips, standardized by precision cutting from media blocks. By using the strips and subdividing them with the aid of diffusion barriers it is

* This Chapter consists of three Sections (I, II and III), variously contributed by Hedén, Illéni and Kühn respectively.

possible to reduce the cost and substantially to simplify biochemical testing. Techniques for encasing and subdividing the strips as well as the equipment used for preparing and for depositing diffusion centres on the inoculated agar surface are described in the first section of this article.

In the second section the choice of an optimal angle for measuring the inhibition and stimulation of microbial growth on an agar surface is described.

Finally, in the third section, the application of the system to the identification of micro-organisms by a battery of biochemical tests is described.

I. INSTRUMENTATION FOR THE BIOCHEMICAL IDENTIFICATION OF MICRO-ORGANISMS*

CARL-GÖRAN HEDÉN

A. Introduction

1. *Limitations of traditional identification methods*

Traditional methods for the identification of micro-organisms depend on a combination of morphological and biochemical characteristics, often supplemented by serology or phage-typing. All those approaches are so difficult to quantify that a reliable result pre-supposes the involvement of an experienced microbiologist. However competent such a specialist might be, his capacity is normally limited by the assistance available for the manual operations. This ranges all the way from dishwashing and media preparation to paperwork, the handling of samples and, of course, the performance of routine laboratory procedures.

Rising costs and the pressure of increasing demands pave the way for rationalization and an avalanche of disposable items, dehydrated media and novel instruments now sweeps over most laboratories. However, the complex and hazardous nature of microbiological work makes a systems approach to the task of identification difficult, and a flexible interphase between the micro-organism and the computer is still a fairly distant goal.

Since many of the major problems which now face mankind have prominent microbiological facets it is important to relieve the qualified microbiologist from routine activities. The same goes for the technical assistants who, particularly in the developing parts of the world, face enormous work loads. Such considerations provided the arguments for UNESCO

* This study as well as the two described in the other sections of this Chapter were supported by grants from NORDFORSK, the Marianne and Marcus Wallenberg Foundation, the Swedish Board for Technical Development, the Medical Research Council and "Försvarsmedicinska forskningsdelegationen".

and WHO to sponsor the first International Symposium on Rapid Methods and Automation in Microbiology (Stockholm, June 3–8, 1973). This aimed at providing an overview of techniques that might help to satisfy the growing needs, among them the so-called "Autoline system" (Hedén, 1974a) described in this Chapter (Hedén, 1974b). The production of prototype units can be arranged by the Karolinska Institutet.

2. *Areas in particular need of automated identification procedures*

The complexity of the microbial populations found in nature is such that the environmental factors which govern important interactions between micro-organisms, man, animals and plants will elude us as long as automated identification methods are not available (Hedén, 1974c). Table I serves as a reminder of some questions in environmental microbiology where such methods might be helpful.

TABLE I

Areas in environmental microbiology where automated identification methods would be helpful:

1. Study of population dynamics in soil and aquatic ecosystems exposed to biocides, fertilizers and other types of environmental pollution.
2. Correlation between the micro-organisms in the digestive tract of man and animals: (a) to the nutritional status of the latter (vitamins, tracemetals, nitrogen balance etc.), (b) to the detoxification of certain food and fodder constituents, and (c) to the production of toxins and carcinogens.
3. Search for micro-organisms with a desirable capacity for selective detoxification of industrial pollutants.
4. Mapping the distribution of strains disseminated from suspected foci such as carriers of disease, trickling filters or microbiological industries.
5. Routine testing of populations and of foods and fodders.
6. Ecological mapping aimed at the development of biological control systems.

The second area of fundamental importance is microbial genetics where the "synthesis" of micro-organisms with desirable metabolic abilities requires a comprehensive description of potential gene donors and gene receptors. In view of the rapid developments in plasmid genetics such descriptions may in fact soon be needed for the legal protection of industrial strains. The replacement of strains deposited in reference collections, when lost is, for instance, a situation where a "metabolic fingerprint" may be valuable. Labour saving devices could also be of great value in relation to the growing science of enzyme engineering where they could facilitate the selection of mutants resistant to catabolic repression, product inhibition and adverse environmental factors (Hedén, 1973; 1974d, e; 1975).

However, the most apparent need is in clinical microbiology, where the routine requirements of virology and for the diagnosis of cancer and auto-immune diseases often place undue pressure on existing facilities.

B. The Autoline system for linear scanning

1. *Basic considerations*

After a careful analysis of the basic needs and engineering alternatives for a flexible mechanization of microbiological methods it was felt that the optical scanning of reactions on or in gels would provide a sound basis for a development programme that might be split into three phases:

(a) Studies of the response of growth films to growth factors and inhibitors. This would not only provide an opportunity to optimize the optical system (Illéni, this Chapter, p. 28) but would also enable a first step to be taken towards the interfacing of metabolic testing and computers.

(b) In the second phase the growth film technique would be adapted to serological and phage typing.

(c) Finally the system would be adapted to pattern recognition of individual colonies derived from mixed populations or cultures exposed to mutagens. The latter might either be used deliberately for strain improvement or provide a target for environmental monitoring (Ames, 1973; Garner, 1972).

The basic hardware that would permit such a development programme has already been described (Hedén, 1974a) and the principal operations are shown in Fig. 1. It is based on the mechanized handling of agar strips

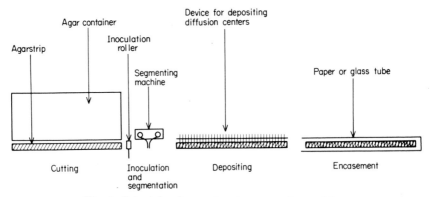

Fig. 1. Principle of operation in agar-strip machine.

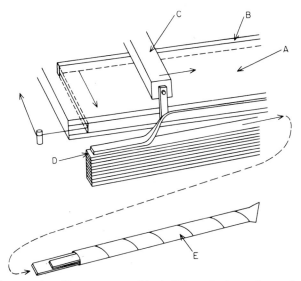

Fig. 2. Top portion illustrates agar block (A) pushed out from glass container by plunger (B). Moving knife (C) cuts off agar strip which is deposited on carrier (D) fed up from below. Finally inoculated agar strip is fed into paper tube (E).

measuring 1 × 50 cm and 2–5 mm in thickness cut from blocks to ensure a degree of reproducibility that would make automation meaningful.

The nutrient-containing agar is poured into a 1 × 50 × 30 cm glass container forming a block. When this is pushed out by a plunger, strips can be cut off by a moving knife (Fig. 2). The strips are deposited on transparent carriers made of glass or plastic fed up from below, and the surface is inoculated by a ceramic roller. Sterility is maintained by intermittent illumination by UV-light and by keeping the critical parts under a sterile air overpressure.

If the agar strip is to be used for a large number of biochemical tests it is automatically transported lengthwise into a second device where reagent ribbons from reels located at right angles to the direction of motion are deposited (Fig. 1). In the course of transport the strip is subdivided into discrete blocks as described below. As a last phase in the sequential operations the cultivation unit (Fig. 3) is fed into a glass or paper tube for incubation.

Automation for numerical taxonomy, has an important prerequisite: the cost of the individual tests must be minimized. To a large extent this is met by the basic concept which permits a reduction both in the labour required and in the quantities of media needed. However, there are two further aspects to be considered:

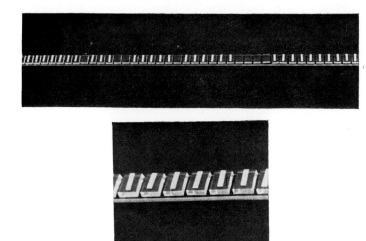

Fig. 3. An agar strip, deposited on a glass carrier, divided into discrete blocks with rayon ribbons carrying the reagents.

(a) using disposable materials for handling the cultures,
(b) the need to establish diffusion barriers in order to make maximum use of the growth surface available.

The solutions to those two problems and their translation into practical devices is the main subject of this section.

2. *Environmental control in a completely disposable system*

The initial approach for handling the agar strips was to carry them on long slabs of glass and to encase them in glass tubes. This permitted mechanized handling of large numbers of samples but, in a fully developed system, would require a complex arrangement for stoppering the tubes and for dishwashing. In view of the need for good optical properties of the carriers used in the initial phase of the studies, these disadvantages were disregarded. In parallel, experiments were performed on a design based on plastic strips. This is a problem of straightforward mechanization which does not merit a detailed description. However, a detailed description of the encasement system is needed. This must take into account:

(a) economy in space and materials.
(b) good heat transfer but efficient containment of humidity and the gaseous environment,
(c) simplicity in closing and opening the encasement.

To meet those requirements a special machine was developed for the *in situ* manufacture of tubes from a reel of three-layered material. This was composed of an inner layer of polythene permitting heat sealing, an intermediary layer of aluminium foil as a diffusion barrier and finally an outer layer of paper permitting simple marking of the tubes. A rough price comparison is given in Table II, and the machine is illustrated in Fig. 4.

TABLE II

Price comparison of construction materials

	Material	Dimensions (mm)	Weight (g)	Price U.S. (cents)
Petri dish	polystyrene	90 diam	~20	~5
*Autoline—tube**	paper (50 g/m²)			
	Alfoil (0·009 mm)	85 × 850		
	polythene (25 g/m²)		~9	~2
strip	plastic (1 mm)	10 × 500		

* Produced by Åkerlund o Rausing, Lund, Sweden.

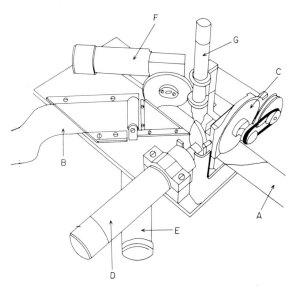

Fig. 4. Paper tube producer. The spiral tube A is made from a three-layered strip B. It is cut into desired lengths by the saw C, which is moved up and down by motor D. Motors E and F move the strip through the machine and position it for welding below the heatsource G.

This also shows the high speed saw which subdivides the endless tube and the thermostatically controlled welding finger which is automatically lifted from the spiral weld between every manufacturing sequence. The device making the transverse welds, closing the tubes, is not shown.

3. *The preparation and deposition of diffusion centres*

Various approaches to the production and deposition of diffusion centres were tested, the main one being based on the use of 2 mm wide strips of fibrous materials impregnated with various inhibitors and growth factors and deposited at intervals on the agar strips at right angles to the long dimension. Figure 5 shows a picture of the device used for subdividing commercially available rayon-fibre bands and Fig. 6 the apparatus used for impregnating the bands. For depositing the diffusion centres at various distances from each other a long guillotine knife is used with a feeding

Fig. 5. Apparatus for subdividing filter paper strips or other types of fibrous material. By means of motor A this is pulled from reel B over a number of parallel highspeed saws operated by motor C.

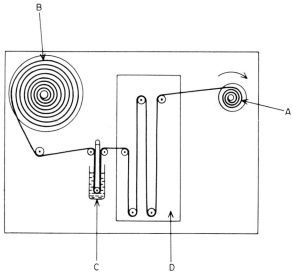

Fig. 6. Apparatus for impregnating filter paper or other types of fibrous material. A motor (A) pulls a fibrous ribbon from a reel (B) through a tube, containing a solution of a chemical (C) and a drying chamber (D).

mechanism, that can be activated individually for each reel. A simplified cross-section is shown in Fig. 7. The device is set up in parallel to the transport route of the inoculated agar strips which can thus simply be provided with a desired combination of chemicals. This is done by activating the corresponding reels. In the case of antibiotic resistance measurements, where the sizes of the zones reflect the sensitivity of the growth film, some ten reels, evenly distributed along the strip, are normally used. In cases where diffusion zone overlaps are desired to disclose synergistic and antagonistic effects, reels are spaced closer together.

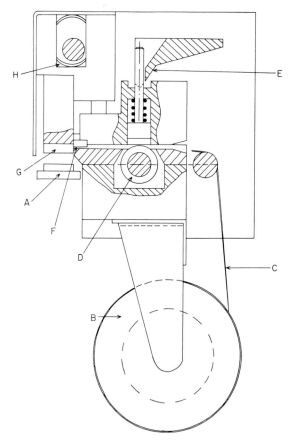

Fig. 7. Cross-section of device for depositing diffusion centres. The gel strip on its glass carrier A is fed from reel B with impregnated material C by means of the motor drive roller D. This moves the material 1 cm at a time provided lever E is in the lifted position. Pieces are cut off between the stationary edge F and the knife G, moved by eccentric disc H.

4. *Alternatives for establishing diffusion barriers*

(a) *The simultaneous use of several basic media.* In microbiology the parallel use of several common media (nutrient agar, blood agar, endo agar, etc.) is standard practice. Since they are all inoculated with the same material it would be desirable to combine them in one surface which could be inoculated in a single operation. In the Autoline system this can be simply achieved by the disposable ceramic rollers used for the inoculations, but it requires the establishment of diffusion barriers in the agar block from which the combined substrate would be cut. This is not an easy problem, because a film barrier might interfere with the cutting process and might also adhere so strongly to the walls of the container that the motion of the agar block might be impeded. If a highly viscous liquid barrier were used it might interfere with microbial growth by being spread over the agar surface in the course of cutting. A gas barrier, finally, would be too unreliable considering the risk of side movements of the individual agar segments.

After a number of tests the method of choice became *in situ* polymerization of an extremely thin polymer barrier. Its nature and use in a comprehensive biochemical identification system is described in the third section of this Chapter (Kühn, this Chapter, p. 36).

(b) *Subdividing the agar strips into discrete media blocks.* For resistance measurements and determining microbiological response to growth factors, the measurement of the dimensions of diffusion zones requires ample space around each diffusion centre. When very large numbers of biochemical tests are to be performed in the course of an accumulation of data

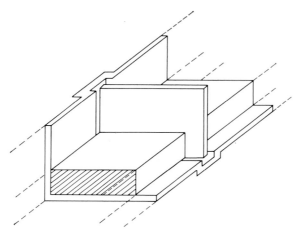

Fig. 8. A segment of a strip which is deposited on a disposable transparent carrier with ditches where the dividers used to segment the inoculated agar are recessed.

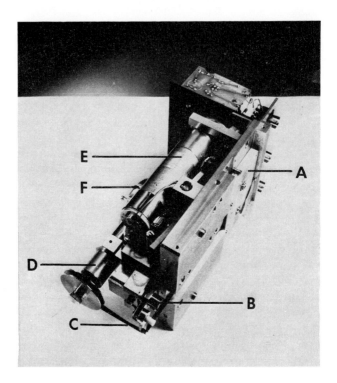

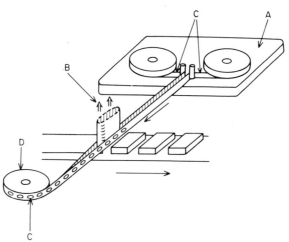

Fig. 9. Segmenting machine. From both reels of standard cassette (A) the two
magnetic tapes (C) are pulled out on both sides of a flattened suction tube (B) by a
small motor (D). The whole suction assembly is moved up and down by another
motor (E) and is connected to a vacuum pump via an air filter (F).

for numerical taxonomy or for obtaining metabolic "fingerprints" of isolated strains, this creates too much "dead space". In such cases diffusion barriers must be established across the inoculated agar strip. Such barriers can be of two types: dividers, made from a disposable material, or ditches, subdividing the agar into discrete blocks. Figure 8 shows a system of simple dividers which make a "clear cut" through the agar, i.e. a cut which leaves no diffusion pathways between the blocks. This is achieved by letting the dividers enter recessed ditches in the L-shaped plastic strip which carries the gel.

For subdividing the agar into blocks the apparatus shown in Fig. 9 was developed. It sequentially lowers a suction tube surrounded by a double knife, made from a disposable tape, over the stepwise advanced agar strip. This device removes 2 mm wide agar segments which are collected in a container that can be easily removed for sterilization. Whenever an agar strip with a new isolate is to be subdivided, a fresh segment of tape is moved into position below the suction tube.

The sequential mode of subdividing the agar strip offers great flexibility in sizing the individual agar blocks and also permits the production of circular sample wells by means of a simple attachment to the suction device. However, this flexibility carries a time penalty and requires a complicated step motor arrangement for ensuring precision in the agar strip positioning. Consequently the simpler "disposable divider" approach is preferable for machines dedicated to numerical taxonomy (Fig. 8). It has been found reliable and adaptable to substance application both on the agar surface and on the bottom side of the gel strip.

REFERENCES

Ames, B. N., Durston, W. E., Yamashi, E., and Lee, F. D. (1973). *Proc. Natl. Acad. Sci.*, **70**, 2281.

Garner, R. C., Miller, E. C., and Miller, J. A. (1972). *Cancer Res.*, **32**, 2058.

Hedén, C.-G. (1973). CIOMS Round table conference: Protection of human rights in the light of scientific and technological progress in biology and medicine. WHO, Geneva, Nov. 14–16.

Hedén, C.-G. (1974a). "New Approaches to the Identification of Micro-organisms", pp. 13–37 (Eds.: C.-G. Hedén and T. Illéni). John Wiley and Sons, New York.

Hedén, C.-G. (1974b). "Instrumentation for the Biochemical Identification of Micro-organisms". Abbreviated version of this paper, read at the Nordforsk Symposium on Automation in Microbiology, Copenhagen, Oct. 1–2.

Hedén, C.-G. (1974c). "The Great Challenge: Applied Microbial Ecology and Some Instrumentation Needs Which it Generates". Introductory paper at 1st Intersectional Congress of the International Association of Microbiological Societies. Tokyo, Sept. 1–7. Introductory volume of proceedings, Science Council of Japan, 1975, p. 5.

Hedén, C.-G. (1974d). "Socio-economic and ethical implications of enzyme

Engineering". Int. Fed. of Institutes for Adv. Study (IFIAS), Stockholm, March. pp. 157.

Hedén, C.-G. (1974e). "1973 Henniker Delphi Study". *In* "Enzyme Engineering. Volume 2", pp. 9–14 (Eds.: E. K. Pye and L. B. Wingard, Jr), Plenum Press, New York.

Hedén, C.-G. (1975). "Enzyme Engineering and the Anatomy of Equilibrium Technology". Symp. on Biophysics and Global Problems, Fifth Intern. Biophys. Congr. Copenhagen, Aug. 8th, 1975. Quarterly Rev. Biophys. (In press).

II. OPTICAL ANALYSIS OF MICROBIAL GROWTH ON TRANSPARENT SOLID MEDIA

TIBOR ILLÉNI

A. Introduction

The use of long agar gel strips (Hedén, 1974 and Hedén, this Chapter, p. 16) instead of conventional Petri dishes is advantageous because great accuracy and reproducibility can be attained. Mechanization of the handling of the strips, for instance their cutting, inoculation and environmental control (temperature, gas phase, etc.) is possible and invites automation. This permits the handling of large numbers of samples but also calls for procedures for automated reading.

Optical methods are commonly used in microbiology (measurement of absorption, transmission, reflection, light scattering, fluorescence, luminescence, etc.) (this Series, Vol. 1). They may be very sensitive and rapid, and can often be used without interfering with the sterility of the object studied. Also photoelectric sensing devices produce millivolt signals which are suitable for computerized data acquisition systems.

In this article scattered light measurements are emphasized because of their potential for yielding information about the numbers, size and shape of microcolonies growing on solid gel surfaces (Illéni, 1973).

B. Principles of the measurement

The aim was to develop a sensitive method for recording density variations in microbial growth developing on nutrient agar gels. The fact that these are presented in the form of strips means great freedom in combining scanning with a variety of light paths.

The possibility of varying the angle of illumination and/or the angle of sensing is naturally important in measurements of scattered light. The principal positions for light sources and detectors around a gel strip-glass slide system are shown in Fig. 10. The illuminator and the light detector are focused on the microbial growth which is limited to the upper surface of the gel strip. Section A is at right angles to the long axis of the

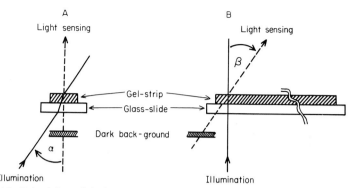

Fig. 10. Principles of dark field illumination of gel strips. A: Illumination from oblique angle α. B: Light sensing from oblique angle β.

gel strip, while Section B is parallel to it. The light source and the detector can be moved, relative to each other in these planes, and in this way various types of darkfield measurements could be made in a search for optimal conditions. However, in practice, since side illumination was very advantageous, only the light source was moved in plane A and the detector in plane B. Measurements using reflected light yielded less information, so illumination angles larger than $90°$ were not considered. Angles for scattered light detection larger than $90°$ have little interest since the scattered light would then have to pass through both glass- and gel layers, causing an unacceptable loss of intensity.

C. The measuring system

On the basis of the principles mentioned above the measuring instrument, shown in Fig. 11, was built and used for scanning the long gel strips.

Gel strips deposited on glass slides ($500 \times 15 \times 2$ mm) are placed on the track of the transport system (T). The transport system which is driven by motor (M) moves the gel strip at constant speed (10 mm/sec) through the measuring zone which is normally about 1 mm^2 in size.

The gel strips are illuminated by a normal microscope lamp (L), (6 V, 15 W) fed by a stabilized power source (ST).* There is a filter holder (F) on the lamp-housing, permitting the introduction of colour filters and polarizers.

The lamp can be rotated in plane A (see Fig. 10) around an optical centre in the middle of the upper surface of the gel strip, and the angle of illumination, α, can be read off directly. The starting point of the scale $\alpha = 0°$, indicates a vertical position of the lamp, i.e. transmission.

* RAC PAC, B32-10R, Oltronix AB, 16229 Vällingby, Sweden.

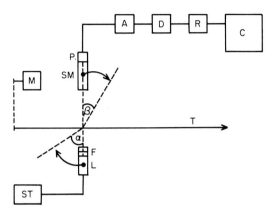

Fig. 11. Schematic diagram of the scattered light instrument. M: Drive motor, T: Transmission track for agar strips, L: Lamp (6 V, 15 W), F: Filter assembly, ST: Stabilized power source, SM: Microscope, P: Photosensor, A: Amplifier, D: Damping, R: Recorder, C: Computer.

The microcolony growth film on the gel surface can be studied by means of a stereo microscope* (SM), at a magnification as low as 2.5×10. A photodiode† (P) mounted on one of the oculars of the microscope is illuminated by the light coming from a 2×0.5 mm spot on the gel surface. Since the measuring principle used is of the dark-field type, there is a black surface just under the glass-gel system.

The stereo microscope and the photosensor can be moved in plane B (see Fig. 10) and the angle to the horizontal, β can be read off directly. The starting point of this scale, $\beta = 0°$, corresponds to the vertical position. The millivolt signals from the photosensor are amplified (A), damped (D) and recorded (R) (see Fig. 11).

The chart-drive motor of the recorder is synchronized with the motor (M) of the strip transport system. This makes possible repeated measurements on the same gel strip, for instance to reproduce a signal from an interesting colony needing more careful study.

In parallel with the recorder a computer‡ can take over and store the signals on a magnetic tape.§ Signals stored in this way can be used for different calculations, e.g. in numerical taxonomy routines (Gyllenberg, 1974 and Kühn, this Chapter, p. 43).

* NIFE MICRAL, Jungner AB, Stockholm 14, Sweden.
† SCHOTTKY BARRIER PIN-10, United Detector Technology, Santa Monica, California 90404.
‡ HP-2100A, Hewlett-Packard Sverige AB, 16120 Bromma, Sweden.
§ HP 7970A, Hewlett-Packard Sverige AB.

D. The optical study of microcolony development

During overnight incubation but sometimes after periods as short as
3 h, a very homogenous microcolony film develops on the upper surface of
the gel strip, provided that this is evenly inoculated, for instance by means
of disposable ceramic rollers (Hedén, this Chapter, Section I). This even
film of microcolonies may be influenced by different reagents, e.g. growth
factors like amino-acids, sugars, vitamins, etc., and inhibitors like antibiotics, antimetabolites and dyes carried by paper strips (2 × 10 mm) that
can be placed on the gel strip after inoculation. Inhibition and/or growth
zones can then be recorded as shown on Fig. 12. Curve A illustrates the

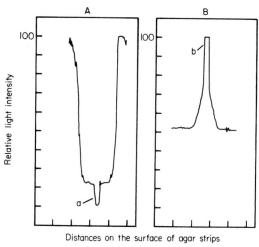

Fig. 12. Typical scattered light responses on (A) inhibition, and (B) growth
zones (a, b: sites of diffusion centres).

effect of the antibiotic Cephalosporine on a growth film of *Micrococcus
luteus* (ATCC 9341), whereas curve B represents a growth zone of *Leuconostoc mesenteroides* (ATCC 8042) around a diffusion centre carrying
lysine, on a lysine deficient medium. The Figure shows that the scattered
light pattern is quite different for the two types of response.

The antibiotic inhibition zone shown in Fig. 12 was measured at $\beta = 0°$
with an optimized angle of illumination $\alpha = 65°$. Optimization was made
with the scanning instrument shown in Fig. 11. The scattered light
intensity was recorded at different angles of illumination, but on the same
inhibition zone. The results are shown in Fig. 13, which illustrates the
dramatic effect of the angle variation. Looking at the curves it is not very
difficult to select the angle at which the contrast between growth and no-
growth areas (signal to noise ratio) is maximal.

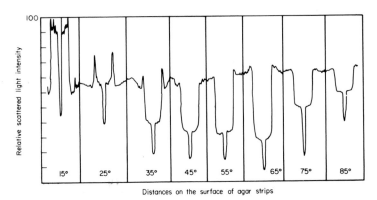

Fig. 13. Scattered light signals as a function of the angle of illumination (α) measured on antibiotic inhibition zones.

Total reflection explains the effects just described (Fig. 14). At an illumination of the strips from large angles, like $\alpha = 65°$, most of the light enters the gel from the side and is reflected on its upper surface. The result is a good dark-field illumination of the colonies, and a reduced scattering from gel turbidities, since they are below the reflecting surface.

At small illumination angles, like $\alpha = 15°$, the scattered light signals are more complicated and less suitable for data processing. At angles like $\alpha = 80°$ the contrast is very low.

At angles about $\alpha = 25°$ the scattered light signals give maxima at the edge of the inhibition zone. This is a very interesting phenomenon, and suggests that at this angle of illumination the instrument is more sensitive to thin growth than to heavy growth. The details of this correlation were studied on a colony-density gradient, that would "expand" the sharp edges

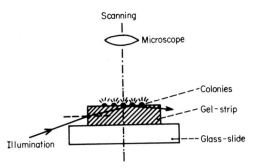

Fig. 14. Total internal reflection in the gel with side illumination. The scattered light from microcolonies can be measured with minimal disturbance from gel turbidities.

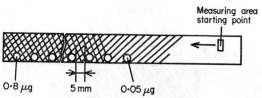

Fig. 15. The performance of microcolony size- and frequency-gradient of *Leuconostoc mesenteroides* (ATCC 8042). The micro-organisms were seeded on to lysine deficient citrate agar strips (3 × 10 × 480 mm), immediately thereafter Munktel S-314 paper discs (5 mm in diameter) charged with increasing quantities of lysine (0·05–0·10–0·15– . . . –0·70–0·75–0·80 γg per disc) were placed upon the surface of the strip. Microcolony sizes and frequencies were measured after 18 h incubation.

of the growth zone. The preparation of a microcolony size and frequency gradient is shown in Fig. 15.

Paper discs were impregnated with L-lysine solutions of increasing concentrations. *Leuconostoc mesenteroides* (ATCC 8042) was inoculated on a L-lysine deficient medium. After incubation this yielded an increasing colony density. When smaller angles of illumination were used disturbances, caused by the edge of the glass slide and gel strip, appeared. Therefore a perpendicular illumination at $\alpha = 0°$ was used and kept constant, while the sensing angle (β) was optimized. The results of the measurements are shown in Fig. 16. The scattered light intensity curve rises more rapidly with increasing colony density at around $\beta = 30°$. This means that for recording low colony densities—e.g. at the edge of a growth zone—an angle $\beta = 30°$ should be chosen. A likely explanation for the phenomenon is that an enhancement at increasing visual contrast angles (as illustrated by Fig. 17) is counteracted by a decreasing level of scattered light on the upper surface of the gel strip. These opposing phenomena would be expressed in the optimum value for β.

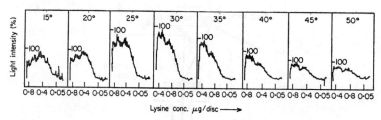

Fig. 16. Relative scattered light intensities as a function of the angle of scanning (β). Measurement on a microcolony size- and frequency-gradient of *Leuconostoc mesenteroides* (ATCC 8042).

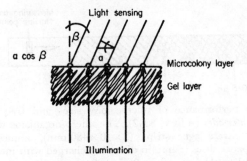

Fig. 17. Principle of visual contrast enhancement by oblique angle light sensing position. No growth : growth quotient on the interval "a" is reduced by cos β.

A further amplification of the scattered light may be achieved by the use of the ring mirror system illustrated in Fig. 18. This collects all the light scattered in a cone defined by β_{opt}. The microscope shown in Fig. 18 is only for focusing the system.

In summary then, oblique angles for illumination should be used for growth inhibition measurements, i.e. for indicating contrast between "heavy growth" and "no-growth". On the other hand the oblique angle sensor position is to be preferred for growth factor analysis, i.e. for measuring the contrast between "very scarce growth" and "no-growth". In the second case total reflection has no significance, since even at optimal conditions, the angle of scattering is lower than that necessary for total reflection on the upper surface of the gel strip.

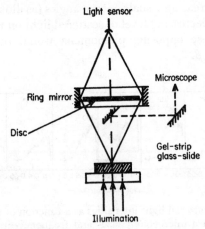

Fig. 18. Signal amplification by ring-mirror.

E. Signal analysis

In Fig. 19 the two typical scattered light intensity curves (A and B) are presented again, together with the damped signals A′ and B′ which are suitable for automated zone diameter reading. The measurement of inhibition zones can be made with high accuracy, because of the great contrast in the scattered light signals from growth and no-growth areas. The measurement of growth zone diameters is more difficult, because the contrast between scarce growth and no-growth is small. However, electronic damping of the signals improves the situation by cutting off the scattered light signals from the areas which are unaffected (Fig. 19, A′ and B′).

Because the gel strips are transported with continuous speed through the scanner, the zone diameter can be measured by a timing device. The intervals are easily recognized and can be corrected for the influence of diffusion speed before being printed via the computer.

The first successful application of the scanning instrument was for the automation of antibiotic sensitivity testing (Wretlind, 1974) but a more general application is for the recording of biochemical tests used in numerical taxonomy. In this case pure cultures of micro-organisms are seeded on a segmented agar medium. Each agar block represents an isolated microbiological reaction to various inhibitors and growth factors, and the scanning instrument, shown in Fig. 11, normally only needs to indicate growth or no-growth. This is a simpler task for the sensor, than detecting zone sizes. In practice, the utilization of total reflection by means of side illumination was found to be the most useful in the scanning of segmented

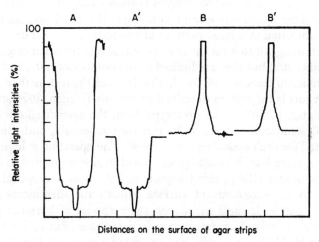

Fig. 19. Typical scattered light responses on inhibition (A) and on growth zones (B), and the same after damping (A′, B′) for automated zone diameter reading.

agar strips, since it offered the best opportunities for reducing the disturbing reflections caused by the diffusion barriers.

The method, used for mechanized biochemical identification work is described in detail in the subsequent section (Kühn, this Chapter, p. 36).

REFERENCES

Gyllenberg, H. G. (1974). *In* "New Approaches to the Identification of Microorganisms", pp. 201–223 (Eds. C.-G. Hedén and T. Illéni), John Wiley & Sons, New York.
Hedén, C.-G. (1974). *In* "New Approaches to the Identification of Microorganisms", pp. 13–17 (Eds. C.-G. Hedén and T. Illéni), John Wiley & Sons, New York.
Illéni, T., Goertz, G., Hedén, C.-G., and Wegstedt, L. (1973). *In* "Abstracts of Papers", B72, pp. 48. Symposium on Rapid Methods and Automation in Microbiology, Stockholm.
Wretlind, B., Goertz, G., Illéni, T., Lundell, S., and Seger, B. (1974). *In* "Automation in Microbiology and Immunology", pp. 293–304 (Eds. C.-G. Hedén and T. Illéni), John Wiley & Sons, New York.

III. DEVELOPMENT OF TESTS FOR MECHANIZED NUMERICAL TAXONOMY

INGER KÜHN

A. Introduction

Identification work in microbiology is normally carried out manually. This has limited the use of numerical taxonomy in this field, because it requires the determination of a large number of properties. Various disposable kits have been designed to meet the needs of rapid identification methods (API, Enterotube, etc.) but they are limited in the number of tests, and flexibility in the choice of tests is very small. The basis of the test method described in this report is the system described previously (Hedén, 1974 and Hedén, this Chapter, p. 16). In its prototype form the system allows groups of about 50 standardized tests to be run simultaneously, and the tests can be selected for each identification purpose. The operation is highly mechanized and evaluation is made by an optical device connected to a recorder and/or computer (Illéni, this Chapter, p. 28). Since the optical system is adapted to measurements of surface growth the identification method described below depends mainly on the response of bacterial growth to different chemicals. The project was initiated as a joint co-operation with Professor H. Gyllenberg, who has presented a more detailed discussion of the potential of numerical taxonomy elsewhere (Gyllenberg *et al.*, 1975.)

Identification of micro-organisms is a large and complex field ranging from placing a bacterium isolated from a clinical specimen into a particular species, to the typing of a strain, belonging to an earlier recognized species. For each identification purpose the most suitable substrate, test battery and cultivation conditions must be selected.

B. Basic principles

1. *Media*

The medium must be adapted to the particular group of micro-organisms and, ideally, all the biochemical tests required for an identification are then carried out on that medium. Two different approaches can be followed:

(a) A medium is used which contains all essential growth factors and permits growth even of quite fastidious micro-organisms. However, it should not allow growth which is too vigorous, since the response to different chemicals then becomes less clearcut.

(b) The other approach is to use a chemically defined medium which lacks some substance necessary for bacterial growth. This can be a carbon source, as Stainer *et al.* (1966) have described for the taxonomy of *Pseudomonas*, or it can be other essential growth factors as in the NEDA-substrate used for auxotyping of Gonococci (Wesley-Catlin, 1973). Both these substrates have also been used with the mechanized system described here.

Table III shows some media used for different identification purposes.

2. *Preparation of agar blocks*

The melted agar is poured into a glass container (see Fig. 2, p. 19). After cooling the agar is cut and inoculated with a bacterial suspension

TABLE III
Different media used in the identification procedures

Medium	Micro-organisms
Standard II-agar (MERCK)	*Enterobacteriaceae*
	Pseudomonas
	Staphylococci
	Enterococci
Mineral salt-medium	*Pseudomonas*
Mineral salt-medium supplemented with CH_3OH	Methanol-oxidizing bacteria
NEDA-medium	Auxotyping of Gonococci

with the aid of a ceramic roller. The agar blocks can also be stored for several weeks in a refrigerator if not used at once. Composite strips composed of two or more different media may also be used. This is done by partly closing the opening of the container which is held vertically while pouring the melted agar. If only a part of the container is filled, new layers can be poured as soon as the previous one has solidified. To prevent diffusion between the layers barriers must however be applied. A mixture of cyanoacrylate and paraffin is used for the purpose. The cyanoacrylate polymerizes very rapidly in contact with a moist surface and produces a polymer film. Paraffin has to be added since cyanoacrylate shrinks while polymerizing. Application is performed as follows:

0·1 ml cyanoacrylate (Aron Alpha Produced by Toa Gosei Chemical Industry Co., Ltd., Tokyo, or EASTMAN 910, produced by Eastman Kodak Co., USA) + 0·1 ml paraffin (mp 52–54°C) is dissolved in 1 ml petroleum ether (bp 40–60°C). This solution is enough for 10 cm² agar surface. Occasionally the cyanoacrylate does not dissolve completely, probably due to ageing, and should then not be used.

The solution is carefully spread over the solidified agar surface, the solvent is allowed to evaporate and a new layer of agar can be poured.

3. *Reagent carriers*

(a) The use of ordinary filter paper to carry reagents introduces difficulties, since a long, 2 mm wide impregnated ribbon is easily torn when wet. Instead non-woven rayonfibre (Paratex Vliestoff I/100, produced by Lohmann KG, Fahr/Rhein Germany) is used. It is very strong, even when wet, and easily cut by the device used for depositing the reagent carriers. Proteins used as reagents are not strongly adsorbed on to this material.

(b) Insoluble reagents, such as blood cells, lipids, etc., are added to a gelatin solution which is cooled and dried. The dried film is then cut for deposition on the inoculated surface. When the agar is incubated the gelatin melts and releases its contents on the agar surface.

C. Biochemical tests

1. *Making a species distinction*

Table II shows some tests which have been used, mainly for the family *Enterobacteriaceae*. The tests were performed on a peptone medium (Standard II agar, Merck), and the amounts of the reagents were calculated for this particular medium. Since it already contains most of the essential growth factors, the additional carbon and energy sources had to be added in relatively large amounts in order to get a measureable growth

TABLE IV

An example of various growth factors, antibiotics and other chemicals which can be used in the Autoline system. The table shows the concentration of the solutions impregnated into the reagent carriers deposited on Standard II-agar

GROWTH FACTORS

	mg/ml			mg/ml
A. Carbohydrates			C. other growth factors	
adonitol	500		α-ketoglutaric acid	500
L+arabinose	500		Na-acetate	500
D-arabinose	500		Na-citrate	500
dulcitol	200		Na-pyruvate	300
levulose	500		Na-succinate	500
galactose	500		glucosamine-HCl	300
glucose	500			
m-inositol	300			
cellobiose	300		METABOLIC TESTS	
lactose	300		aesculin	100 mg/ml aesculin
maltose	500			+50 mg/ml $FeCl_3$
mannitol	500			
mannose	500			
melibiose	500		H_2S	200 mg/ml cysteine+
raffinose	500			40 mg/ml $Pb\,(OAc)_2$
rhamnose	500			
salicin	300			
sorbitol	500		INHIBITORS	
ribose	500			mg/ml
sucrose	500		A. Antibiotics	
trehalose	500		Cephalotin	5
xylose	500		Sulfaisodimidin	20
amygdalin	300		Trimetoprim	2
glycerol	500		Erythromycin	2
erythritol	500		Tetracyclin	1
sorbose	500		Chloramphenicol	2
lyxose	200		Streptomycin	2
			Benylpenicillin	10
B. Amino-acids			Kanamycin	2
asparagine	300		Ampicillin	5
cysteine	200		Nitrofurantoin	5
Na-aspartate	300		Nalidixin	2
Na-glutamate	300		Colistin	5
glycine	300		Carbenicillin	5
histidine-HCl	300			
lysine-HCl	500		B. Other inhibitors	
proline	500		NaN_3	20
serine	300		Na-desoxycholate	50
threonine	200		Methyl violet	5
OH-proline	300		$CuSO_4$	20
arginine-HCl	500		$MnSO_4$	100
alanine	200		$CoCl_2$	5
β-alanine	200		$NiCl_2$	10
valine	100		Brilliant green	2
γ-aminobutyric acid	500			
glutamine	200			

stimulation. Only $0.28 \times 2 \times 10$ mm fibre ribbon pieces could be used as reagent carriers for 0.25 cm^3 agar blocks, so relatively large concentrations had to be used for the impregnation. Consequently only substances which were easily soluble in some solvent could be used in this approach. This is a limitation which of course did not cause problems when dried substance was deposited on a transparent agar carrier and permitted to penetrate the agar from below.

For inhibitors the amounts which give total inhibition on blocks inoculated with sensitive bacteria and no inhibition of resistant bacteria are used (see Table IV).

Some tests are used in which reagents for metabolic products are added to the medium. Such tests are common in the differentiation of

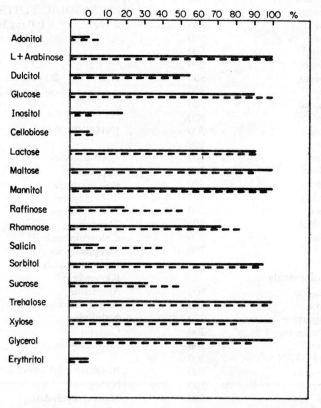

Fig. 20. The figure shows the percentage of *E. coli* strains which give positive results in the presence of the sugars listed. The results were measured by two methods: a (black line) Growth measured by light scattering. 22 strains used. b (dotted line) Determination of acid production. 1231 strains used (Edwards *et al.*, 1972).

Enterobacteriaceae. Of these only the H_2S production and aesculin hydrolysis tests were included since they could be read directly by the optical device. Other tests for extracellular products were rejected, either because they demand special substrates or a more complicated optical system than the one currently used.

The optical system is so sensitive that no advantage was found in using a pH-indicator in the gel. In fact much larger quantities of carbohydrate were required to produce a distinct colour change than those which caused a clearcut stimulation of growth. More sophisticated optical techniques, like a simultaneous measurement at two wavelengths (the dicro-principle introduced for titrations by Ringbom *et al.* (1967)) were tested but not found necessary for ordinary biochemical testing.

Even though the number of strains used was very small, Fig. 20 still indicates that the mechanized system and a standard method for measuring acid production from carbohydrates give fairly similar results.

TABLE V
Various reagents used for differentiation among Methanol-oxidizing bacteria

Acetate	L-Tartarate	γ-Aminobutyric acid
Citrate	D-Tartarate	Threonine
Formate	Aesculin	Sucrose
α-ketoglutarate	β-Alanine	Xylose
Maleinate	Glutamate	β-Methylglucoside
Malonate	Histidine	Acetamide
Pimelinate	Lysine	Urea
Succinate	OH-Proline	Salicin
Suberate	Serine	Methionine

2. Differentiation within a species

Sometimes it is necessary to recognize a special strain of a known species. For such investigations other tests than the ones used for distinguishing at the genus level are used. Table V shows some tests which are used for a preliminary differentiation of methanol-oxidizing bacteria.

D. Gas phase control

Since the incubation is performed in closed tubes there are certain matters which must be considered. It is for instance important to select tests which do not release volatile metabolites that might influence the growth on neighbouring blocks. Such interactions of course also exclude the use of volatile reagents. However, such drawbacks can probably be compensated by the very large number of new tests that can be introduced and by the high level of standardization that can be achieved when exactly the same test battery configuration is always used. In principle it should also be possible to use dividers which separate the gas phases over different tests.

E. Experimental

A 3 mm thick, 10 mm wide and 500 mm long strip of agar is cut from a gel block and inoculated with a disposable ceramic roller dipped in a bacterial suspension. The density of the inoculum should be relatively high ($T_{640} \sim 80\%$) in order to yield an even microcolony surface. The strip is then automatically subdivided into 8 mm long blocks which are charged with 10×2 mm rayonbands, carrying the chemicals, selected for the test (Fig. 21). The strip is fed into a tube which is closed and incubated.

After incubation the growth is measured by the method described earlier (Illéni, this Chapter, p. 28) at an illumination angle of 65°. The results are either visualized on a recorder (Fig. 22) or put on a magnetic tape for computer analysis.

In Fig. 22 some control blocks can be seen. They are essential whenever the bacteria can grow without added reagents, since they provide a baseline reading between inhibition and stimulation.

If a greater discrimination between the results is desired, the surface under each peak can be measured. This is very useful, for instance, in cases where one wants to establish identity between isolates, belonging to the same species.

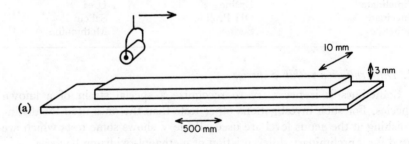

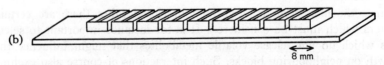

Fig. 21. An agar strip. (a) Before preparation; (b) ready for incubation.

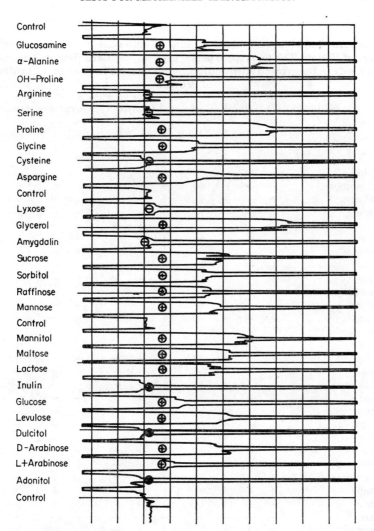

Fig. 22. The figure shows the recorder response of *Klebsiella pneumoniae* grown on Standard II-agar to which was added the various reagents indicated.

F. Discussion and illustrative result

1. *Numerical taxonomy*

The classification of micro-organisms by means of numerical taxonomy is based upon the determination of a large number of characters of different strains and the subsequent calculation of similarities. In conventional diagnostic work just a few characteristics are measured and some of them

TABLE VI

The results from some typical strains of different bacteria run with a battery of 40 tests. The tests were performed on Standard II-agar. The results are coded: + for growth stimulation, − no change in growth, ○ inhibition of growth

	E. coli				Klebsiella			Ps. aeruginosa		Enterococci	
	1	2	3	4	5	6	7	8	9	10	11
Mannitol	+	+	+	+	+	+	+	+	+	+	+
Adonitol	−	−	−	−	+	−	+	−	−	−	−
L+Arabinose	+	+	−	+	+	+	+	−	+	−	−
Levulose	+	+	+	+	+	+	+	+	+	+	+
Galactose	+	+	+	+	+	+	+	−	−	+	−
Glucose	+	+	+	+	+	+	+	+	+	+	+
Inositol	+	−	+	−	+	−	−	−	−	−	−
Cellobiose	−	−	−	−	+	+	+	−	−	+	+
Lactose	+	+	+	+	+	+	+	−	−	+	−
Maltose	+	+	+	+	+	+	+	+	−	+	+
Mannose	+	+	+	+	+	+	+	+	+	+	+
Melibiose	+	+	+	+	+	+	+	−	−	−	−
Raffinose	−	−	+	+	+	+	+	−	−	−	−
Rhamnose	+	+	+	−	+	+	+	−	−	−	−
Salicin	−	−	−	−	−	−	−	−	−	+	+
Sorbitol	+	+	+	+	+	+	−	−	−	+	−
Sucrose	−	−	+	−	+	+	+	−	−	−	−
Trehalose	+	+	+	+	−	+	+	−	−	+	+
Xylose	+	+	+	+	+	+	+	−	−	−	−
Sorbose	+	+	+	−	+	−	−	−	−	−	−
Dulcitol	−	+	+	−	−	−	−	−	−	−	−
Proline	+	+	+	+	+	+	+	+	+	−	−
Serine	+	+	+	+	+	+	+	−	−	−	−
Alanine	+	+	+	+	+	+	+	+	+	−	−
Glutaminate	+	−	+	−	+	+	+	+	+	−	−
Cysteine	+	+	+	+	−	+	+	+	+	−	−
Acetate	+	+	+	+	+	+	+	+	+	−	−
Citrate	−	−	−	−	+	−	+	+	−	−	−
Pyruvate	+	+	+	+	+	+	+	+	+	−	−
Succinate	+	+	+	+	+	+	+	+	+	−	−
Glucose-NH$_2$	+	+	+	+	+	+	+	+	+	+	+
Aesculin	−	−	−	−	+	+	+	−	−	+	+
H$_2$S	+	+	+	+	+	+	+	−	−	−	−
Sulpher	−	−	−	−	−	○	○	−	−	−	−
Erythromycin	−	−	−	−	−	−	−	−	−	−	−
Ampicillin	○	−	○	○	○	○	○	−	−	−	−
Colistin	○	○	○	○	○	○	○	−	○	−	−
Na-desoxycholate	−	−	−	−	−	−	−	−	−	−	−
Brilliant green	○	○	−	○	○	−	−	−	−	−	−
Chrystal violet	−	−	−	−	○	−	−	−	−	−	−

TABLE VII
Similarity index

Isolate no.

1	2	3	4	5	6	7	8	9	10	11	
100	90	88	90	78	80	73	60	65	48	40	1
	100	83	90	68	75	68	60	65	53	45	2
		100	78	75	83	75	58	58	45	38	3
			100	68	80	73	65	70	58	50	4
				100	78	80	48	48	40	33	5
					100	93	55	60	53	45	6
						100	58	58	45	43	7
							100	90	63	70	8
								100	58	65	9
									100	93	10
										100	11

are regarded as more important than others. In the numerical approach on the other hand every measured character may be given the same weight. A large number of properties must however be determined, and a recommendation is to use a minimum of 60 tests (Sneath *et al.*, 1973). To demonstrate one of the simpler approaches to numerical taxonomy, described by Lockhard *et al.* (1970), some actual experiments will be shown below.

Table VI gives the results obtained with 40 tests on some typical strains. The tests were selected because of their similarity to those often used in diagnostic work concerned with the family *Enterobacteriaceae*. Since the number of calculations increases with the square of the number of strains studied, the results with only 11 strains are used in this illustrative experiment. They were obtained from the clinical laboratories where they had first been classified.

The results for each isolate are compared with those obtained with all other isolates and a similarity coefficient, *s*, is calculated. This is, in this case, defined as the number of tests which give the same result in a compared pair, divided by the total number of tests employed. In Table VII the results of these calculations are shown. Each value refers to a pair of isolates listed in Table VI. In Table VIII the pairs are listed in order of decreasing *s*-values.

The next step is to combine the isolates into groups where the members are linked together at the highest similarity level of that group, as shown in Table IX. From Table VIII the pair or pairs with the highest *s*-value are picked—in this case 6–7 and 10–11. Since they have no member in

TABLE VIII
**Pairs of isolates from Table V, arranged in order
of decreasing S values**

% S	Individual pairs
93	6–7, 10–11
90	1–2, 1–4, 2–4, 8–9
88	1–3
85	
83	2–3, 3–6
80	4–6, 1–6, 5–7
78	1–5, 3–4, 5–6
75	2–6, 3–7, 3–5
73	1–7, 4–7
70	4–9, 8–11
68	2–7, 2–5, 4–5
65	9–11, 1–9, 4–8, 2–9
63	8–10
60	1–8, 2–8, 6–9
58	3–8, 3–9, 4–10, 7–8, 7–9, 9–10
55	6–8
53	6–10, 2–10
50	4–11
48	1–10, 5–8, 4–9
45	6–11, 7–10, 3–10, 2–11
43	7–11
40	1–11, 5–10
38	3–11
35	
33	5–11

TABLE IX
Isolates arranged into groups

% S	Groups
⩾93	(6, 7) (10, 11)
⩾90	(6, 7) (10, 11) (1, 2, 4) (8, 9)
⩾88	(6, 7) (10, 11) (1, 2, 4, 3) (8, 9)
⩾83	(6, 7, 1, 2, 4, 3) (10, 11) (8, 9)
⩾80	(6, 7, 1, 2, 4, 3, 5) (10, 11) (8, 9)
⩾70	(6, 7, 1, 2, 4, 3, 5, 8, 9, 10, 11)

common they make up two different groups, which is indicated by the parentheses. The sorting is continued with the next s-value, including also the groups from the previous row which may now be expanded by other pairs provided that they have a member in common with the previous group. At 90 % similarity the pairs 1–4 and 2–4 have 4 in common, and consequently make up one group (1, 2, 4). The pair 8–9 on the other hand makes another separate group. At the next s-value, 88 %, 1 and 3 make a pair and therefore belong to the same group as 1, 2, 4, where it is added at the end. This sorting is finished when all individuals end up in the same group.

The results can then be visualized in many ways, for example, in the form of a dendrogram (Fig. 23). In this lines combining the isolates are then drawn at their highest s-value in Table VII.

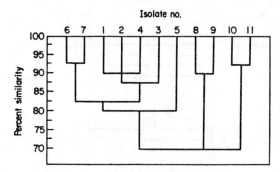

Fig. 23. A dendrogram constructed from the data in Table VII.

In Figure 23 four different groups can be seen, containing *Klebsiella*, *E. coli*, Enterococci and *Pseudomonas*, and one "Outlier" (Niemelä *et al.*, 1974), which had been named *Klebsiella* in the clinical laboratory. The example shows that even with such a simple battery of tests as those used in this experiment, the results do not differ much from those obtained with conventional methods.

Needless to say the analysis requires a computer and, since this can easily be programmed also for determining the discrimination potential of various tests, these can be modified or replaced as more and more previously classified strains are processed.

Figure 24 shows the results of a larger-scale experiment carried out with the same tests as in the previous example. Thirty-eight strains mainly belonging to the families of *Enterobacteriaceae* and *Pseudomonodaceae* were tested. This time the growth intensity as height of the peaks was measured and the results were coded from 1 to 5, where 1 stands for inhibition of

growth and 5 for very good growth. The results were analysed according
to the methods described by Niemelä *et al.* (1974).

The strains were divided into four groups by the analysis. In group 1
all *E. coli* strains were found together with strain no. 35 which had been
classified as *Pseudomonas. Klebsiella* and *Enterobacter* were not separated
by the test battery used and made up group 2. Group 3 was rather inhomo-

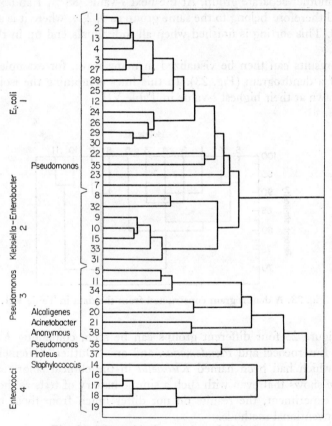

Fig. 24. Hierarchical grouping of 38 isolates.

genous, containing mainly *Pseudomonas,* but since the tests were selected
for members of the family *Enterobacteriaceae* this is not surprising. The
enterococci make up a group which is quite far from the Gram-negative
rods.

Figure 25 finally records the results from an experiment to differentiate
among various isolates of *E. coli.* Those numbered 2, 3, 9, 4, 1 and 8
were received from the Finnish National Public Health Laboratory and

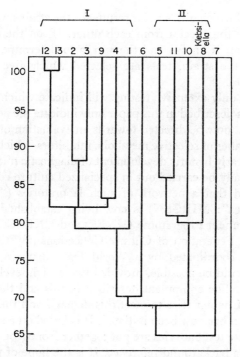

Fig. 25. Grouping of presumptive *E. coli* strains.

the rest from the United Nations Ethiopean Coli Project -74. They were suspected of producing a strong enterotoxin.* The figure, which was prepared according to the methods described above, shows two distinct groups of bacteria, one containing all the Finnish *E. coli* isolates and two Ethiopian, the other group containing a typical *Klebsiella* and 3 of the Ethiopian *E. coli* isolates. The other isolates are "outliers". According to this figure, the Ethiopian isolates nr. 5, 11, 10 are actually not *E. coli*, but rather *Klebsiella*, a conclusion that was confirmed by later findings. An interesting point in the figure is that isolates nr. 12 and 13 exhibited 100 % similarity in spite of the fact that they came from different patients. Such a similarity indicates that they belong to the same strain. This conclusion was also confirmed by a closer investigation.

G. Conclusions

The final formulation of a test system for automated numerical taxonomy must be determined by the aims of the work and by the kinds of bacteria concerned. If the aim is to differentiate between species of a family,

* Wadström *et al.*, to be published.

tests are selected which give similar results for the strains within the species but discriminate the species from each other. If, on the other hand, the species of a strain is known and the aim is to determine the "metabolic fingerprint" of the strain, tests are selected which give varying results within the species.

In both cases only extensive testing will indicate which tests should be used, so the ones described in this paper only indicate the general approach. Attention should now be directed towards an evaluation of novel tools like the growing number of selective metabolic inhibitors, which were not available during the early historic development of diagnostic microbiology. This work should ideally be carried out in specialized culture collections, and it should be noted that a network of such laboratories ("Microbiological Resource Centres": MIRCEN) is now being established by the United Nations Environment Programme (UNEP) and UNESCO in cooperation with the World Federation of Culture Collections (WFCC). Since those centres will be coordinated by a World Data Bank in Brisbane rapid development should be possible, provided that a high level of standardization is achieved. The equipment described in this and the preceding two sections of this Chapter tries to meet this demand and the production of a simplified version has now been initiated. It is based on a modular concept where appropriate test batteries are put together from groups which consist of 10 tests plus a background reference. It is the hope of the authors that this effort will help to generate more specialized kits of the disposable type which has helped to simplify diagnostic microbiology so much. Far from being a competitor with simple methods the automated test procedures may in fact be a prerequisite for their rapid development.

REFERENCES

Edwards, P. R., Ewing, W. H. (1972). "Identification of Enterobacteriaceae". Third edition. Burgess Publishing Company.

Gyllenberg, H. G., and Niemelä, T. K. (1975). *In* "New Approaches to the Identification of Micro-organisms", pp. 201–224.

Hedén, C.-G. (1974). *In* "New Approaches to the Identification of Micro-organisms", pp. 13–37 (Eds. C.-G. Hedén and T. Illéni), John Wiley & Sons, New York.

Lockhart, W. R., and Liston, J. (1970). "Methods for Numerical Taxonomy". American Society for Microbiology.

Niemelä, T. K., and Gyllenberg, H. G. (1974). Reports from Department of Microbiology, University of Helsinki, **8**.

Ringbom, A., Skrifvars, B., and Still, E. (1967). *Anal. Chem.*, **39**, 1217–1221.

Sneath, P. H. A., and Sokal, R. R. (1973). "Numerical Taxonomy", W. H. Freeman and Company, San Francisco.

Stainer, R. Y., Palleroni, N. J., and Doudroff, M. (1966). *J. Gen. Microbiol.*, **43**, 159–271.

Wesley-Catlin, B. (1973). *J. Infec. Dis.*, **128**, 178–193.

CHAPTER III

Gas–Liquid Chromatographic Chemotaxonomy

D. B. DRUCKER

Department of Bacteriology and Virology,
University of Manchester, Manchester, England

I. INTRODUCTION

A. Gas–liquid chromatographic chemotaxonomy

Gas–liquid chromatographic chemotaxonomy (GLC-chemotaxonomy), as both a term and a technique, has only recently appeared in the micro-biological literature. Taxonomy, the theoretical study of classification (Sokal and Sneath, 1963) and chemotaxonomy, the use of chemical analytical data in taxonomy, and classification, are terms which will be readily understood. Traditional chemotaxonomy employs such well known techniques as cell wall analysis (Cummins and Harris, 1956) DNA base ratios (Hill, 1966) and polyacrylamide gel electrophoresis of protein extracts (Kersters and DeLey, 1975).

In this Chapter GLC-chemotaxonomy will embrace not merely chemical analyses used in taxonomy but also GLC-based microbial detection and identification, which are not aspects of taxonomy, *strictu sensu*, at all.

The gas chromatograph is now seen much more commonly in micro-biology laboratories than would have been the case only a decade ago; however, the principles of its operation are sufficiently veiled in mystery to many microbiologists to warrant the following brief description of the apparatus and terms used in gas chromatography.

B. Gas chromatography

1. *Background and principles*

The term "chromatography" is attributed to the Russian biochemist Tswett who successfully separated plant pigments using a column of a solid adsorbent. This technique of liquid–solid chromatography (LSC) was extended by Martin and Synge (1941) who used an elution technique with solid-bound liquid as a stationary phase; their technique was an example of liquid–liquid partition chromatography (LLC). The use of a gas to replace the liquid moving phase was suggested by the same authors and in 1952, James and Martin reported in the literature their work on gas–liquid partition chromatography, also called vapour phase chromatography (VPC), gas–liquid chromatography (GLC) and gas-chromatography (GC), although the latter term embraces an older technique of gas–solid chroma-tography (GSC). The classical paper of James and Martin (1952) described the simultaneous separation of several fatty acids, which represented a tremendous advance over the then available techniques. For GLC analysis to be possible, the sample to be analysed must be volatile, if only to a limited extent, or to be convertible to a volatile derivative, or degradable to volatile products. The technique can now be used to examine bacterial fatty acids, sugars, amino-acids, purine and pyrimidine bases. By scaling up the analytical technique, preparative GLC is practicable.

GLC offers the chemical microbiologist two advantages—(i) extremely low concentrations for detection of most substances, and (ii) rapid analysis. It also has the disadvantage that absolute identification is rarely possible without resorting to the use of ancillary techniques.

A full treatment of GLC theory is outside the scope of this Chapter and the reader is referred to the standard texts which exist (Littlewood, 1970); the following is intended merely as a brief outline of some theoretical and practical aspects of interest to the microbiologist. Under conditions of linear ideal chromatography, molecules of the substance to be analysed after injection on to a gas chromatographic column would instantly partition between the moving gas phase, e.g. a stream of nitrogen and the stationary liquid phase absorbed on to the column packing material in accordance with the partition coefficient, $K = \dfrac{[\text{solute}]_{\text{liquid}}}{[\text{solute}]_{\text{vapour}}}$. The passage of "carrier" gas through the column would remove molecules in the vapour phase and molecules in the liquid phase would instantaneously leave the liquid phase to preserve the value of K. Molecules in the vapour phase would not diffuse backwards against the flow of carrier gas. The column would be uniformly packed so as to maintain a constant proportion of gas : liquid. A mixture of two substances A and B injected on to such a column, and having differing values of K would be separated and be eluted at separate instants of time, called the "retention time"; frequently the volume of carrier gas required to elute the substance is referred to ("retention volume").

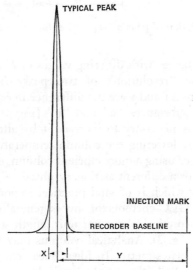

Fig. 1. Diagrammatic GLC peak. Key: (X) peak width, (Y) retention time.

In practice, linear non-ideal chromatography is observed in which substances are not eluted at single instants of time but show "band-broadening" such that if the concentration of eluted substance is plotted *vs* time a characteristic asymmetric peak is obtained (Fig. 1). This is because of the following reasons:

    (i) no column is ever packed absolutely uniformly,
   (ii) the measured value of K is not constant for numerous reasons such as diffusion of vapour molecules, diffusion of solute molecules,
  (iii) equilibrium is not instantaneously reached.

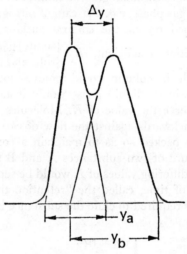

FIG. 2. Calculation of peak's "separation factor" $= 2\Delta y/(ya+yb)$.

Therefore, two substances with differing values of K may not be satisfactorily resolved. The "resolution" of two peaks (separation factor) $= 2[\Delta y/(y_a+y_b)]$ where Δy and y are the difference in retention volumes and the peak widths of substances "a" and "b" (Fig. 2). If the separation factor is too low it is necessary to increase it by altering the chromatographic conditions by lowering the column temperature, or reducing the carrier gas flow rate, or using a more efficient column, e.g. a longer column, a repacked column, or a different stationary phase.

Column efficiency which is of vital practical importance to the microbiologist engaged in GLC-chemotaxonomy is generally measured in terms of the number of theoretical plates per unit length where the number of plates, $n = 16(x/y)^2$ (Fig. 3). Analytical columns may be expected to have approximately 2000 plates/metre. In ideal linear chromatography $n = \infty$. As an alternative to "n", the height equivalent to a theoretical plate

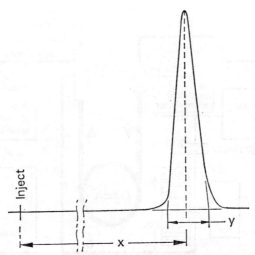

FIG. 3. Calculation of number of theoretical plates $n = 16(x/y)^2$.

(HETP) is frequently referred to; where, HETP $= l/n$ and $l =$ length of column in mm.

2. *The gas chromatograph*

As with any sophisticated piece of apparatus, failure to use the gas chromatograph correctly will produce results—but results which will possibly be quite useless. Some understanding of the instrument is thus essential: the gas chromatograph is not a single instrument but an assortment of modules, which may be interchanged almost at will depending upon the type of analysis required. The various modules will be considered in turn below. For detailed information on the operation of gas chromatographs the reader is referred to the various manufacturers' handbooks.

Figure 4 shows a diagrammatic representation of the gas chromatograph, in which a carrier gas is passed through pressure-, or flow-, controllers to the column. A suitable injection system injects a sample into the beginning of the column, which is maintained at a predetermined temperature, and substances eluted from the column pass from the far end of the column through a detector system. An electric current from the detector is amplified and fed to a suitable recording device, usually a chart-recorder, possibly via an integrator, or incorporating an integrator.

(a) *Gas supplies.* The carrier gas is usually oxygen-free nitrogen; it may be high purity argon or helium. The latter gas is frequently used in the U.S.A. where it is more readily available than in the U.K. Argon is generally used in conjunction with the electron-mobility detector and the electron capture detector. Because of the high sensitivity of the gas chromatograph, traces of

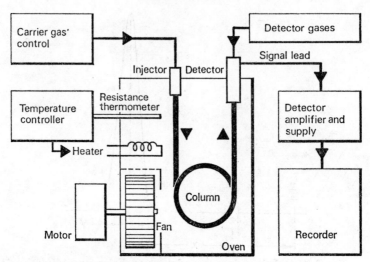

FIG. 4. Diagrammatic representation of gas chromatograph.

organic impurities in the carrier gas may be detectable and create problems. The use of purifying bottles containing a molecular sieve is most advisable. With some of the more commonly used detectors, additional gas supplies are required, viz., hydrogen and air. With the latter two gases, the use of purifying bottles is absolutely essential, otherwise noise spikes will be created by impurities in the gases and trigger an electronic integrator, if used, or create a poor chart-recorder base-line. Temporary alleviation is possibly by judiciously flaming and tapping the H_2 and air lines.

In the simplest variants of gas chromatograph, the gas flow rates are adjusted by altering the gauge-head outlet pressure; more complicated instruments include flow controllers. Gas flow rates used vary between laboratories. Typical flow rates for 5 ft analytical columns are: $N_2 = 45$ ml/min; $H_2 \times 40$ ml/min; air $\times 650$ ml/min. Incorrect H_2/air ratios can cause alteration in detector sensitivity and at low operating temperatures cause condensation problems in the detector, if the instrument used has no detector oven.

(b) *Sample injection.* Samples can be injected on to the column as solid, in glass vials; as liquid, by syringe; or as gas, by gas sampling valve, depending on the nature of the substance to be analysed. In chemotaxonomy, solutions are generally injected by a syringe of 1–10 μl capacity, through a silicone rubber septum. A common practical difficulty is septum leakage which results in a falling base-line on the chart-recorder and delayed retention times for peaks.

Since immediate equilibrium between liquid and vapour phase is strived for, injection ports frequently use an injection point heater which raises the injection point temperature above that of the column and ensures rapid evaporation of the sample. Syringes must always be well flushed with a solvent after use.

(c) *Columns.* These may be either stainless steel or glass, of various dimensions, and either packed or open tubular (capillary). A typical packed analytical column may be 2 m in length and of 4 mm i.d. tubing. A typical capillary column may be 25 m in length and of 0·25 mm i.d. tubing. Capillary columns utilize an "injection splitter" to reduce sample size and prevent overloading. In order to fit the chromatograph oven, columns are "u"- or "w"-shaped, or coiled.

In capillary columns the stationary phase is simply coated on the inside of the column. In packed columns, it is first coated on to an inert support material.

Ideally, supports should, (i) have a specific area of 1 m^2/g, (ii) be chemically inert, (iii) have thermal and mechanical stability, and (iv) consist of uniformly sized particles. Two of the most commonly used supports are crushed firebrick and diatomaceous earth. Neither material is chemically inert and therefore can react with polar solutes, as a result of surface-OH groups which can be replaced by a trimethylsilyl group after silanization. Ready-treated supports are available.

For certain separations, polymers are used in place of a support coated with stationary phase, e.g. Poropak, used for fermentation products (*vide infra*); this is an application of gas-solid-chromatography.

The choice of stationary phase is critical and radically affects the separation attempted. A few of the hundreds of stationary phases available are shown in Table 1. Generally speaking, the polarity of the stationary phase must be suited to the polarity of the sample.

Columns can be purchased ready packed and it is advisable for the novice to buy a ready-packed column. However, coating the support and packing a column are fairly simple and economical operations. A suitable stationary phase should be:

(i) virtually non-volatile at the analysis temperature (which should be approximately 50°C below the solute's boiling point at 760 mm Hg),
(ii) thermally stable,
(iii) have a suitable value for K,
(iv) be chemically pure,
(v) be inert towards the solute.

The stationary phase is normally applied to the support after solution in a suitable organic solvent (Table I) and excess solvent removed with a rotary evaporator. Care must be taken to prevent a "ball-mill effect" in which the

TABLE I
Characteristics of some stationary phases commonly used in chemotaxonomy

Stationary liquid	Abbreviation	Approximate max. operating temperature	Support coated with solution in	Application
Methyl silicone gum	SE30	300°C	CH_2Cl_2	TMS derivatives; carboxylic esters
OV1-silicone	OV1	350°C	$CHCl_3$	TMS derivatives; carboxylic esters
Apiezon L	APL	260°C	$(C_2H_5)_2O$	Esters
Dexsil	—	500°C	$CHCl_3$	Amine derivatives
Free fatty acid phase	FFAP	275°C	$CHCl_3$	Separation of carboxylic esters including enoate classes
Polyethylene glycol adipate	PEGA	195°C	CH_2Cl_2	
Diethylene glycol adipate	DEGA	195°C	CH_2Cl_2	
Diethylene glycol succinate	DEGS	185°C	CH_2Cl_2	
Ethylene glycol adipate	EGA	180°C	CH_2Cl_2	
Ethylene glycol succinate	EGS	185°C	CH_2Cl_2	
Neopentylglycol succinate	NGS	195°C	CH_2Cl_2	
Polyethylene glycol 400	PEG 400	85°C	CH_2Cl_2	Polar compounds; alcohols, esters, ketones. With alkali, amines
Polyethylene glycol 20M	PEG 20M	220°C	CH_2Cl_2	

support is damaged. Normally supports are coated with 1–10% w/w of stationary phase. Increasing proportions of stationary phase reduce the "tailing" effect, which is seen as non-gaussian peaks, but increase retention volumes.

Metal columns are packed then coiled; glass columns are precoiled. To assist even packing, it is usual to apply a vacuum to one end of the column and to rest the column against a vibrator, e.g. a "Rotamix". Coiled columns should be packed so that the packing is always travelling "downhill".

Generally, packed columns are preheated for 24 h at the maximum operating temperature plus 20°C unless this is above the limit for the stationary phase. Carrier gas is passed through the column at twice the "normal" flow rate. Volatile impurities are driven off, which would otherwise create a high background signal. Sometimes improved resolution is obtained with reduced preheating.

All columns "bleed", even after preheating, so resulting in a small background signal which increases if the operating temperature is raised. After column preheating, some packing-down may have occurred, in which case the column is "topped-up" with fresh packing before being plugged with glass wool.

In the case of the Chromosorb 101 used for fermentation analysis, it is advantageous to carry out preheating prior to column packing (personal communication from Professor Jan Carlsson).

(d) *Oven and temperature controls.* Increase in temperature not only results in increased "bleed" as noted above but also results in a reduction in retention volume. The former is seen as a rising chart-recorder base line: the latter is noticed as peaks appear "too soon", when they possibly risk being mis-identified.

For isothermal GLC, the temperature is very strictly controlled and since it can take approximately 2 h for column temperature equilibration, it is advisable to leave the gas chromatograph running overnight. This also restores column efficiency.

For certain chemotaxonomic applications, e.g. pyrolysis of bacterial cells or amino-acid analysis, the substances to be analysed differ considerably in K value and under isothermal conditions the more volatile substances appear as closely-bunched peaks whilst the less volatile substances may have low, broad peaks which appear very much later.

To overcome this problem, temperature-programming is possible, when the oven temperature is raised at a pre-determined rate. Increased column bleed is compensated for by means of a second "blank" column whose increased bleed results in increased detector current which is opposed to that of the "test" column. This results in the bleed signal being nullified.

For many chemotaxonomic applications, workers have used inlet injection heaters to cause rapid vaporization of samples and better peak resolution. Sometimes, also, a detector heater has been used with advantage.

(e) *Detector systems.* The ideal detector must (i) be able to detect minute quantities of resolved sample, (ii) have a linear response up to six decades, (iii) have equal response to different classes of compounds, (iv) have rapid response, (v) possess stability, simplicity, low noise.

Various types of detector are available. The construction of some of the commoner variants is shown in Fig. 5.

The detector most commonly used for chemotaxonomic studies is the flame ionization detector (F.I.D.) in which eluted molecules form ions upon combustion in an air–hydrogen flame. The air/H_2 flow rates must be carefully determined. This detector can measure nanogram quantities of most substances except for a few compounds including, unfortunately, formic acid which is a common fermentation product of interest in GLC-chemotaxonomy.

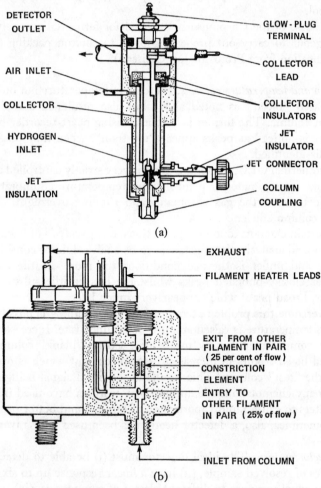

(a)

(b)

FIG. 5. Some common GLC detectors: (a) flame ionization, (b) thermal conductivity.

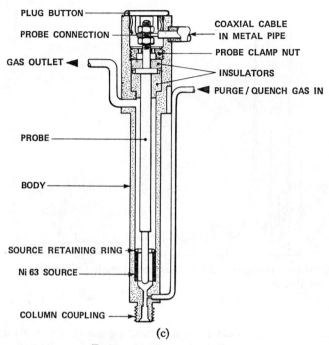

FIG. 5. (c) Electron capture.

The thermal conductivity cell or Katharometer (T.C.D.) can detect formic acid but only measures μg amounts of substances; it measures changes in thermal conductivity resulting from elution of sample components.

The electron capture detector (E.C.D.) is tremendously sensitive to compounds containing electronegative groups or atoms. In chemotaxonomy, fluorinated compounds including derivatives of bacterial metabolites and co-metabolites have been examined using this technique.

(f) *Amplifiers.* These fulfil a number of functions; they supply the detector potential, permit "backing off" or electrical zeroing of the background detector signal created by column bleed and allow attenuation of the strength of the signal coming from the detector.

The cheaper amplifiers are linear over only three decades and "saturate" with large solvent peaks. This phenomenon, which is seen as either flat-topped peaks or split peaks, renders them unsuitable for use with electronic integrators, which require an amplifier having linearity of response over six decades, and having low drift characteristics.

The amplifier output may be fed to a chart recorder, directly, or via an electronic integrator. Integrators are discussed below in Section VI.

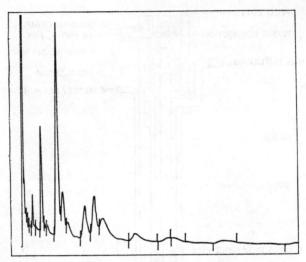

Fig. 6. Streptococcal fatty acid chromatogram; methyl carboxylic esters separated at 190°C on PEGA.

(g) *Chart recorders.* These trace a chromatogram of the analysis (Fig. 6) which provides the experienced gas chromatographer with information on sample size, retention characteristics, possible overlap of peaks, tailing, column efficiency and peak resolution.

For GLC work, recorders should preferably be of the flat-bed type so that information can be written on the chromatogram, and should have a series of inputs to match the various possible amplifier and integrator outputs. They should have variable chart speed and rapid response time with a good stylus which will not "run dry" on the ascending limb of early peaks. Some recorders are combined chart recorders and disc-integrators (see Section VI).

C. Application of GLC to chemotaxonomy

In view of the versatility of the gas chromatograph, with regard to the range of compounds that can be analysed, it is not entirely surprising that GLC has been used as a chemotaxonomic tool in numerous ways.

The main GLC-chemotaxonomic methods have been:

(i) analysis of fermentation end-products,
(ii) analysis of structural and other components of microbial cells,
(iii) pyrolysis gas–liquid chromatography,
(iv) analysis of body fluids for detection of micro-organisms.

1. Analysis of fermentation end-products

This type of analysis is generally less useful in the case of aerobes and more fruitful when applied to anaerobic organisms which tend to use organic compounds as terminal electron acceptors, so reducing them to taxonomically interesting acids, alcohols, and ketones.

Pioneers in the application of fermentation analysis to identification and taxonomy were Orla-Jensen (1919) who proposed setting up taxonomic groups on biochemical rather than morphological grounds, and Prévot (1940) whose "Manuel de Classification des anaerobies" lists fermentation end-products of many bacterial species.

Although production of propionic acid has been a key identification characteristic of members of the propionibacteria for many years, only recently with the advent of GLC has its simple and routine detection been practicable. Facultative anaerobes, also are amenable to fermentation analysis, which is of some importance in the lactic acid bacteria. Table II lists some typical fermentation end-products, based on the identification schemes of the "Anaerobe Laboratory Manual" (Holdeman and Moore,

TABLE II

Characteristic fermentation products

Representative micro-organisms	End-products of fermentation
Yeasts	Ethanol, carbon dioxide
Lactic acid bacteria	
(homolactic)	Lactic acid; traces of formic and acetic acids, ethanol
(heterolactic)	Lactic and acetic acids, ethanol and carbon dioxide, some formic acid
Propionibacterium	Propionic and acetic acids, carbon dioxide
Escherichia	Acetic, formic and lactic acids; ethanol, carbon dioxide, hydrogen and trimethylene glycol
Klebsiella	Acetylmethylcarbinol, butan-2,3-diol, diacetyl *plus* end products of *Escherichia*
Clostridium	Very variable, according to species. Usually butyric acid and smaller amounts of other acids, alcohols, ketones, etc.
Leptotrichia	Lactic acid; traces of acetic, formic and succinic acids
Fusobacterium	Butyric acid; variable amounts of lactic, acetic, propionic, formic and succinic acids
Butyrivibrio	Formic acid; sometimes also lactic acid; traces of butyric, acetic and succinic acids
Bacteroides	Usually succinic acid and acetic acid, occasionally also lactic acid; traces of formic propionic, *iso*-butyric, *iso*-valeric, butyric, and valeric acids

1972). Products of interest include formic, acetic, propionic, n-butyric, *iso*-butyric, n-valeric, *iso*-valeric, lactic, and succinic acids; ethanol propan-1-ol, propan-2-ol, and pentan-2-ol; butan-2,3-diol, diacetyl, acetone, acetoin, hydrogen and various organic amines; the latter arising chiefly by decarboxylation of α-amino-acids (Lambert and Moss, 1973).

2. *GLC analysis of cellular composition*

This technique was first suggested by Abel *et al.* (1963) who examined carboxylic methylesters derived from members of the *Bacillaceae, Enterobacteriaceae, Micrococcaceae* and *Parvobacteriaceae* and concluded that "gas chromatograph analysis of chemical composition has potential feasibility as a method for the rapid classification of bacteria" and accurately predicted that analysis of amino-acid or carbohydrate composition would also prove of value.

This technique has generally involved analysis of volatile derivatives such as trimethyl silyl ethers or methyl esters of cellular components such as sugars and fatty acids, or amino-acids (Moss *et al.*, 1971a). Most genera have now been subjected to this type of analysis which is at least as useful when applied to aerobes and facultative anaerobes as it is with obligate anaerobes. Methyl esters of fatty acids of a large range of micro-organisms have now been studied, including *Nocardia, Escherichia coli, Mycobacterium, Corynebacterium, Bifidobacterium, Lactobacillus, Clostridium, Streptococcus, Mycoplasma, Neisseria, Serratia, Agrobacterium,* treponemes, *Pseudomonas, Moraxella, Micrococcus, Bacillus, Salmonella, Streptomyces,* extreme

TABLE III
Characteristic cellular fatty acids of bacteria

Family	Characteristic fatty acids	
	Major acids	Minor acids
Pseudomonodaceae	$C_{16:0}$, $C_{16:1}$, $C_{18:1}$	$C_{14:0}$, $C_{18:0}$, $C_{18:2}$
Bacillaceae (Bacillus)	br-$C_{14:0}$, br-$C_{15:0}$, br-$C_{16:0}$, $C_{16:0}$, br-$C_{17:0}$	$C_{14:0}$, $C_{15:0}$, br-$C_{16:1}$
Lactobacillaceae	$C_{16:0}$, $C_{16:1}$, $C_{18:1}$, cyc-$C_{19:0}$	$C_{14:0}$, $C_{18:0}$, cyc-$C_{17:0}$
Micrococcaceae	br-$C_{15:0}$, $C_{16:0}$, $C_{16:1}$, br-$C_{17:0}$	$C_{15:0}$, $C_{15:1}$
Corynebacteraceae	br-$C_{15:0}$, $C_{16:0}$, br-$C_{17:0}$, $C_{18:0}$	$C_{14:0}$, $C_{15:0}$, $C_{16:1}$, $C_{17:0}$, $C_{18:1}$, $C_{21:0}$, $C_{22:0}$, plus extra long chain acids
Enterobacteriaceae	$C_{16:0}$, $C_{16:1}$, cyc-$C_{17:0}$, $C_{18:1}$, cyc-$C_{19:0}$	$C_{14:0}$

$br = iso$-, and/or, *anteiso*-acids; cyc = cyclopropane acids.

thermophiles, rumen bacteria, and blue–green algae. Typical analyses, of fatty acids, appear in Table III.

Only 11 years before the classical paper of Abel and co-workers, James and Martin (1952) had described the technique of gas–liquid chromatography, using the homologous series of fatty acids to try out their analytical method; thus, it was only natural that Abel should favour the analysis of cellular fatty acids using well-tried methods.

Latterly, the development of very good GLC separations for sugars and amino-acids has stimulated interest in the analysis of these cellular components. Since structural components, although being "turned-over" metabolically, are less "environment-dependent" than fermentation end-products, their analysis seems highly desirable.

Recent work has attempted to overcome the inherent problems of subjectivity and long data analysis times by applying statistics and computers to microbial chemotaxonomy, based on fatty acid methyl ester analysis (Drucker, 1974).

Although GLC fingerprinting does not require absolute chemical identification, it is, possible to make absolute identifications after the application of ancillary techniques.

3. *Analysis of pyrolysates*

Pyrolysis–gas liquid chromatography (PGLC) was originally applied to micro-organisms by Reiner (1965) and Oyama and Carle (1967) and has now been applied to a range of micro-organisms including *Mycobacterium* (Reiner *et al.*, 1971), *Aeromonas* and *Vibrio* (Haddadin *et al.*, 1973), *Aspergillus* (Vincent and Kulik, 1970), *Clostridium* (Cone and Lechowich, 1970) and *Streptococcus* (Huis, In't Veld *et al.*, 1973). Comparison of pyrograms obtained by different workers is not generally possible, of course, because even tentative peak identifications are generally unknown and experimental conditions have not been internationally standardized.

One interesting application of PGLC is its use in the U.S. space programme for detection of possible microbial life on the planet Mars!

The advantage of PGLC is the extreme simplicity of the methods used and the rapidity of analysis; the disadvantage is that the "fingerprint" or "signature" obtained cannot meaningfully provide information on the chemical structure of the pre-pyrolysed cell.

The actual techniques used in PGLC are described in Section IV of this Chapter.

4. *Analysis of body fluids*

This is microbial detection and identification, rather than microbial taxonomy. Methods currently available can detect certain micro-organisms chemically at lower population densities than could be detected bacterio-logically; since with an electron capture detector, the gas chromatograph can detect pg quantities of certain metabolites in serum, cerebro-spinal fluid, or synovial fluid.

Mitruka and Alexander (1969) devised a novel method in which cometa-bolism of halogenated organic acids by organisms present in low concen-tration resulted in the production of electrophore-containing substances detectable at extremely low concentration. Such techniques have been found to increase detector sensitivity to *E. coli* up to 20,000-fold. More recently, the same works showed that various *Clostridium* spp. can co-metabolize trichloro-phenoxyacetate to products detectable in minute amount by the ECD.

II. TECHNIQUES FOR GLC ANALYSIS OF FERMENTATION PRODUCTS

A. Analysis of acidic end-products of fermentation

Although the introduction of GLC revolutionized the analysis of many acidic fermentation end-products, which had formerly been analysed by fractional distillation, colorimetry, paper and liquid chromatography, problems are still posed by the diversity of acid products. These range from the monocarboxylic acids separated by James and Martin (1952) to keto-acids, hydroxy-acids and carboxylic acids. These substances represent-ing different homologous series, have widely differing volatilities and thermal stabilities, so that their separation on one column at the same time creates methodological problems for the microbiologist. Additional problems are the failure of the FID to detect formic acid and the "tailing" of acidic compounds on many columns, owing to interactions of the polar molecules with their own species and with the support.

A number of workers have attempted to overcome some of these prob-lems by preparing more volatile derivatives. This approach overcomes three problems—(i) lack of volatility of polyfunctional acids, (ii) lack of detectability of formic acid by FID (which can, however, detect formate esters) and (iii) acidic tailing, but creates a new problem, viz., loss of extremely volatile formate and acetate esters.

Another solution to the problem of lack of volatility of lactic acid has been its GLC analysis following oxidation to acetaldehyde by periodate on a pre-column, heated above oven temperature by an injection point heater. Unfortunately, in the author's opinion, this method cannot be recommended as the reaction is obviously not specific for lactic acid. Formation of lactones by lactic acid standard is yet another problem, as is the failure of GLC to distinguish between stereoisomers of lactic acid, which may have chemo-taxonomic significance.

Most methods described in the literature are complicated by steps to remove microbial cells, from which acetic and other acids might otherwise be leached. Other complications are procedures to remove protein by the addition of denaturants, or procedures to remove the acids from the other components of the supernatant, by ether extraction or by the use of ion-exchange resins. Ether extraction of supernatants brings its own crop of problems: (i) many acids are preferentially soluble in water, (ii) their partition coefficients are not constant and vary with molarity of acid present and with the presence of other acids, (iii) pH of the supernatant will drastically affect the proportion of acid present as undissociated acid.

Despite all these practical problems which the microbiologist should be aware of, many elegant techniques are now available for the detection of acid fermentation products and will be discussed under the following headings: Culture of organisms, GLC analysis of free acids, GLC analysis of acid esters, GSC analysis of free acids and esters.

1. Culture of organisms

No standard medium is universally employed. The composition of the medium is generally dictated by the nutritional requirements of the organism which may be met by a semi-synthetic medium (Henis *et al.*, 1966), a complex medium (Carlsson, 1973) or even a blood agar plate (Brooks *et al.*, 1971) (Table IV). Liquid media have been used both for batch culture and continuous culture of micro-organisms.

The most widely used cultural conditions are probably those laid down by the Virginia Polytechnic Institute (V.P.I.) anaerobe laboratory. Their method employs a pre-reduced peptone–yeast extract–glucose medium based on their basal medium.

V.P.I. Peptone–yeast extract–glucose medium

Peptone	1·0 g
Yeast extract	1·0 g
Resazurin solution	0·4 g
Salt solution	4·0 g
Distilled water	100·0 g
Cysteine-HCl.H$_2$O	0·05 g

(Holdeman and Moore, 1972)

For the more fastidious organisms they recommend supplementation with bile, Tween-80, vitamin K, heme, serum, or rumen-fluid.

The medium used should be analysed before being used for culture, since most media contain acetic acid and other substances *before* microbial growth has taken place.

TABLE IV

Fermentation acid analytical parameters (simplified)

Author	Growth of organism	Treatment	Sample analysed	GLC column	Temperature	Detector
Brooks et al., 1971	5% Rabbit blood agar 36°C/24 h	Acidification; hot CHCl₃ extrn or ether extrn Butylation	Butyl esters	3% OV-1	80°C for 1 min/ to 265°C at 7.5°C/min	Dual F.I.D.
				3% TCEPE and Resoflex LAC 1-R-296	100°C for 5 min/ to 150°C at 5°C/min	
Holdeman & Moore, 1972		Acidification ether extn dry with MgSO₄ or methylation, CHCl₃ extrn	Free acids / Methyl esters	Resoflex 1-R-296	118°C	T.C.D.
Henis et al., 1966	Semi-defined medium, standard inoculum, 35°C/40 h	Acidification, addition of Na₂SO₄, ether extrn. drying of extract	Free acid	10% Carbowax 4000 terminated with terephthalic acid	110°C / 70°C	F.I.D. / E.C.D.
Drucker, 1970	Brain heat infusion, 37°C/48 h	Acidification, ether extraction methylation of sodium salts	Free acid / Methyl† esters	10% DEGA on phosphoric acid treated celite	120°C	F.I.D.
Carlsson, 1973	Holdeman & Moore's PYG up to 5 days	Drainage through cation-exchange resin / Methylation and drainage through anion-exchange resin	Free acid / Methyl esters	Chromosorb 101	68°C / 200°C	F.I.D. / Hydrogen-flame detector

TCEPE = tetracyanoethylated pentaerythritol.

† PEG 400 more suitable

The qualitative analysis of fermentation products is not particularly dependent on normal growth conditions; although for quantitative analyses, experimental parameters need to be carefully standardized.

One point frequently not appreciated is that "characteristic" fermentation acids are not produced in equimolar amounts; for example a *Clostridium* sp. which produces butyric acid as a characteristic fermentation acid, may produce it at a molarity many times higher than the propionic acid which is a characteristic fermentation end product of *Propionibacterium* spp. thus requiring a smaller culture volume for analysis.

2. *GLC analysis of fermentation acids*

Preparation of free fatty acid solutions from culture supernatants generally involves acidification and ether extraction. Water dissolved in the diethyl ether may, or may not, be removed by deep freezing or by the addition of anhydrous sodium, or magnesium, sulphate using a variety of procedures (Table IV). An alternative is the use of ion-exchange resins (Carlsson, 1973).

Ether extraction has the disadvantage that not only fermentation acids but hydrophobic substances in general are extracted and injected on to the column of the gas chromatograph. This difficulty is circumvented by an additional step which consists of washing the ether extract with an alkaline solution, followed by re-acidification and re-extraction.

A suitable technique is the following:

To 9 ml culture supernatant is added 1 ml 2N H_2SO_4 (with or without an internal standard consisting of *n*-caproic, or *n*-heptanoic, acid). Add 1 volume A.R. diethyl ether and shake vigorously, using a standardized shaking routine. This extraction need not be carried out in a separating funnel; in the author's experience, a clean "universal" bottle is satisfactory. If an emulsion forms, a sharp tap on the extraction vessel will usually cause the emulsion to "break". Alternatively, light centrifugation of the "universal" bottle and its contents is possible. Retain the ether extract; add one drop BDH "Universal Indicator"; run in 1N NaOH dropwise with shaking until the acid has been extracted from the ether. Discard the ethereal phase and retain the aqueous phase. Re-acidify with 1 ml 2N H_2SO_4 and re-extract with 1 vol analytical grade diethyl ether. The diethyl ether extract now contains less impurity than formerly and the acids will have been concentrated. The extract will now be ready for injection on to the gas chromatograph column. Suitable columns will generally be acidic to prevent "ghosting" (Geddes and Gilmour, 1970) and one very satisfactory column is 10% w/w diethylene glycol and 2% phosphoric acid on diatomite C (DEGA packing). Separations of acids from C_2-upwards are shown on this column in Fig. 7. Retention data of acid fermentation products are shown in Table V.

The temperature at which the analysis is performed will obviously vary with the nature of the column packing, but with a 10% w/w DEGA

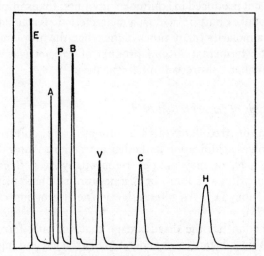

Fig. 7. Gas-chromatogram of n-carboxylic acids from C_2–C_7, separated at 130°C on DAP. Key: (A) acetic, (P) propionic, (B) butyric, (V) valeric, (C) caproic, (H) hexanoic, (E) ether solvent.

TABLE V

Relative retention times of acid end-products of fermentation

	GLC column		
Free acid	DAP at 130°C	DAP at 100°C	Chromosorb 101 at 195°C
Formic†	—	—	0·76
Acetic	*1·00*	*1·00*	*1·00*
Propionic	1·29	1·62	1·65
iso-Butyric	1·59	1·84	2·36
n-Butyric	2·08	2·50	2·78
iso-Valeric	2·44	—	4·12
n-Valeric	3·40	4·90	4·91
iso-Caproic	4·57	—	7·59
n-Caproic	5·49	—	8·80
Lactic‡	—	—	5·81
Succinic‡	—	—	16·68

† Formic acid is not generally regarded as detectable by F.I.D.

‡ Succinic and lactic acids are not generally regarded as "volatile" substances for the purpose of GLC.

column, 120°C is a satisfactory temperature for the separation of free volatile acids.

Because of the corrosive properties of organic acids, it is imperative that the GLC syringe be adequately washed out with an organic solvent after each injection of an ethereal solution of acids.

The detection of formic acid by katharometer is possible, but sensitivity will be poor. To detect formic acid present in trace amounts, esterification is generally advisable so that the F.I.D. can be employed. Typical esterification methods are described below.

3. GLC analysis of esters of fermentation acids

Esterification of free acids is not difficult but does pose some problems. Esters of formic acid will be extremely volatile, e.g. methyl formate, b.pt. 31·8°C (at 760 mmHg); whilst esters of non-volatile acids will be far less volatile, e.g. methyl lactate, b.pt. 144·8°C. The large variety of reagents which have been employed (Table IV) reflects this dilemma.

The production of butyl esters by Brooks *et al.* (1971) results in good recovery of formic and acetic acids but produces less volatile esters of non-volatile acids; e.g., butyl lactate b.pt. 75°C at 6 mmHg. For this reason, methyl esters have been more generally employed, and various methylation reagents are described in Section III of this Chapter. Two methylation techniques of value in fermentation analysis are as follows:

(a) *Method of Holdeman and Moore* (1972). 1 ml acidified culture, 2 ml methanol and 0·4 ml of 50% H_2SO_4 are heated to 55°C in a stoppered vessel for 30 min. Then 1 ml water and 0·5 ml chloroform are added. After gently extracting the methyl esters which have formed, into the chloroform layer, 14 μl of chloroform extract are injected on to the GLC column.

(b) *Method of Drucker* (1970). After having added a suitable internal standard, e.g. *n*-heptanoic acid, to the microbial supernatant, 9 ml supernatant are added to 1 ml 2N H_2SO_4 and 10 ml reagent grade diethyl ether. Fermentation acids are extracted by shaking for standard lengths of time, e.g. for 1 min in 5 min repeated three times. The acids are recovered from the ether extract with sodium hydroxide as described in Section II.A, 2 of this Chapter. The sodium salts can be kept indefinitely after being dried in air.

When required for analysis, the dried salts, in a bijou bottle, are powdered with a microspatula and 1·5 ml 15% w/w boron trifluoride in methanol is added. The vessel is sealed and methylation takes place rapidly and quantitatively at room temperature.

After 40 min the vessel is cooled in a freezer at −18°C for 20 min.

During this time, sodium fluoride precipitates and loss, by evaporation, of methyl formate is reduced to an acceptable level.

Injection of esters on to a 5 ft glass column of 10% w/w DEGA, or 10% w/w PEG 400 on Diatomite C, is carried out swiftly to prevent re-warming, and excessive loss, of volatile esters. This method has the disadvantage that one or more esters are masked by the methanol peak and that insufficient care over cooling can result in loss of methyl formate. It has the advantage of quantification and increase in sensitivity since acids from 9 ml of supernatant are ultimately converted to a methyl ester solution in 1·5 ml methanol reagent.

Since excess boron trifluoride, in the presence of moisture produces hydrogen fluoride, extreme care must be taken over syringe washing, for hydrogen fluoride can corrode both steel and glass and cause seizure of the plunger and erosion of the glass syringe barrel.

Separation of methyl acetate and methanol is much improved on PEG 400 columns, especially if pre-heating of the column is kept to a minimum. The methylating reagent used has a limited shelf life of about 3 months, before it produces spurious peaks (detected by injection of reagent blank) and forms a precipitate of boric acid (due to hydrolysis by atmospheric moisture).

4. *GSC analysis of fermentation acids*

Gas–solid chromatography, until recently, was used mainly for the separation of such gases as the inert gases and methane, oxygen, nitrogen, carbon monoxide, carbon dioxide, oxides of nitrogen and hydrogen. Recently the technique has been applied more generally with the advent of porous polymers, e.g. polystyrene granules. Porous polymers are rapidly being used for separations where GSC has not traditionally been employed. One of these separations is the analysis of microbial fermentation products.

The use of porous polymers PAR-1 (Rogosa and Love, 1968) and Chromosorb 101 (Carlsson, 1973) have enabled fermentation supernatants to be chromatographed directly, after acidification, or after a simple elution through cation-exchange resin.

Carlsson's technique has the possibly unique advantage that simply treated fermentation liquor can be analysed for all the common volatile acids, including formic acid, as well as the non-volatile acids, pyruvic, lactic and succinic, ethanol and other alcohols, diacetyl, acetoin and butan-2, 3-diol (Carlsson, personal communication). Why non-volatile acids or formic acid should produce peaks is incompletely understood.

The Chromosorb 101, 80/100 mesh, employed by Carlsson is best pre-conditioned before column packing at 250°C, and for analysis is run isothermally at 200°C, although the precise conditions can be varied to favour good resolution of particular compounds.

B. GLC analysis of neutral end-products of fermentation

The neutral end-products include representatives of several homologous series of compounds such as alkanols, alkanals, ketones, and certain polyfunctional compounds, e.g. diacetyl, acetoin, and butan-2,3-diol. The chemical diversity of the neutral compounds does mirror, to a lesser extent, the difficulties encountered with free acids.

If ether extraction is employed, for example, diacetyl is much more preferentially soluble in diethyl ether than acetoin would be. Neutral products range in volatility from acetaldehyde (b.pt. 20·2°C) which is readily lost by evaporation to compounds like butan-2,3-diol (b.pt. 193°C).

Separation of certain normal straight chain and branched chain alcohol isomers presents almost insurmountable difficulties.

Many of the growth conditions and column operating conditions, described above, are suitable for the analysis of neutral-end products, especially if slightly lower oven temperatures are used. A summary of techniques used is shown in Table VI.

1. *GLC analysis of alcohols*

The main problem is the separation of ethanol from propan-2-ol. The problem of tailing peaks observed with free acids is not encountered with alcohols so that although the acidic columns, used for free acids, can be used for alcohols, they are not necessary and polyethylene glycol columns are quite adequate. Figure 8 shows a chromatogram of an analysis upon such a column.

If PEG 400 is used as a liquid stationary phase, it is worth remembering the relatively low maximum operating temperature of this material (Table I).

After growth of the test organism and centrifugation, it is possible to analyse the supernatant directly but this is hardly advisable as extraneous material is introduced into the column and artefactual peaks result.

A more satisfactory alternative is to either extract with diethyl ether, which has the disadvantage of poorly extracting lower alcohols, or to pass the sample through an anionic, then a cationic, exchange resin, before analysis. Relative retention data are shown in Table VII. Purification of samples is less critical if a porous polymer GSC method is used.

A typical method for analysis of alcohols is as follows: After addition of *n*-propanol as an internal standard (unless this is already present), culture supernatant (10 ml) is extracted with 1 vol analytical grade diethyl ether,

TABLE VI

Analysis of neutral end-products of fermentation

Author	Growth conditions	Treatment	GLC column	Temperature	Detector	Cpds separated
Carlsson, 1973	Holdeman & Moore's PYG, up to 5 days	Drainage through cationic-exchange resin	Chromosorb 101	200°C	Hydrogen flame detector	Alcohols
Chen & Levin, 1974	Vitamin-free casein hydrolysate for 5 days	Ether-extraction	Carbowax 20M	143°C	—	Phenethyl† alcohol
Doelle, 1967	—	—	30% PEG 400	70°C or 110°C	Hydrogen flame detector	Acetone, alcohols, diacetyl
Doelle, 1969	—	—	30% PEG 400	50°C for 5 min; 10°/min; 120°C	Hydrogen flame detector	Acetone, alcohols, diacetyl
Holdeman, & Moore 1972	Growth in PYG for 5 days	Ether-extraction	Resoflex-R-296	118°C	T.C.D.	Alcohols
Mitruka & Alexander, 1972	Thioglycollate-broth	Ether-extraction	4% SE-30	190°C	F.I.D. E.C.D.	Alcohols, diacetyl, acetoin, butan-2,3-diol
Palo & Ilkova, 1970	Milk	—	Porapak Q & Porapak P	105°C	F.I.D.	Acetone, alcohols, acetaldehyde, diacetyl

† Strictly speaking, not a fermentation product.

using a standardized shaking routine (*vide supra*). After drying of the extract with anhydrous sodium sulphate the solution in diethyl ether will probably require a ten-fold concentration. Standard conditions for evaporation should be used, e.g. using steam or compressed air. Two μl quantities of

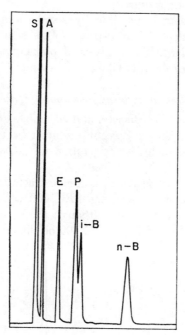

Fig. 8. Gas-chromatogram of some neutral fermentation end-products separated on polyethylene glycol. Key: (S) solvent, (A) acetone, (E) ethanol, (P) propanol, (*i*-B) *iso*-butanol, (*n*-B) *n*-butanol.

TABLE VII

Relative retention times of neutral end-products of fermentation

Neutral compound	DAP at 130°C	DAP at 100°C	Chromosorb 101 at 195°C
Acetone	0·21†	0·12	0·73
Ethanol	0·21	0·14	0·58
iso-Propanol	0·21	0·14	0·76
Diacetyl	0·29	0·18	1·11
n-Propanol	0·29	0·21	0·92
iso-Butanol	0·29	0·21	1·14
n-Butanol	0·41	0·37	1·55

† Retention time is relative to acetic acid.

concentrated extract can be injected, in triplicate, on to a 5 ft column of
10% PEG 400 on diatomite CAW (100–120 mesh) operated at 73°C with a
carrier gas flow rate of 40 ml/min. Peaks are detected by F.I.D. For all the
alcohols of interest, calibration curves should be plotted, unless only
qualitative analyses are required.

Quantitation of data is possible by standard injection techniques or
introduction of an alcohol (e.g. *n*-propanol), absent from the supernatant
as an internal standard. The analysis of phenethyl alcohol produced by
Achromobacter is described by Chen and Levin (1974).

2. *GLC analysis of acetaldehyde, acetone, diacetyl, acetoin and butan-2,3-diol*

The analysis of acetone, diacetyl and acetaldehyde is possible using the
technique described above. Precautions must be taken to avoid loss of
acetaldehyde. These include refrigeration of extracts and use of sealed
extraction vessels.

The substances can be analysed, as ether extracts, using the columns listed
in Table VI. On some columns, e.g. PEG 400, separation of diacetyl and
ethanol may present difficulties, although the author has obtained good
separations on columns of 10% PEG 400 on diatomite CAW at 73°C.

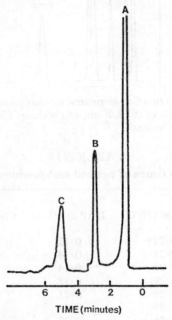

TIME (minutes)

Fig. 9. Gas-chromatogram of ethyl acetate (internal standard) and diacetyl. Key:
(A) ether, (B) ethylacetate, (C) diacetyl (after Lee and Drucker, 1975). Reproduced
by kind permission of the American Society for Microbiology.

The analysis of acetoin and butan-2,3-diol is more difficult as these substances have very low partition coefficients between ether and water, and diols have long retention times on columns operated at temperatures satisfactory for the analysis of the more volatile neutral fermentation end-products.

It is possible to analyse both compounds very satisfactorily by injection of culture supernatants on to Chromosorb 101 at 200°C (Carlsson, personal communication). Microbial acetoin may also be injected on to PEG 400 as an ether extract after preliminary oxidation to diacetyl by the following method (Lee and Drucker, 1975).

Organisms are grown in a glucose-containing medium such as brain–heart infusion for 48 h at 37°C and the cultures centrifuged at +4°C. The supernatants are extracted with 1 vol analytical grade diethyl ether containing 0·2% v/v ethyl acetate as an internal standard by shaking for five 30 sec periods during half an hour.

GLC analysis as described above (Fig. 9) permits quantitation of diacetyl. The majority of the acetoin (if any) will still be in the aqueous phase. It can be analysed after conversion to its α-diketone derivative, diacetyl, by oxidation with 12% w/v ferric chloride hexahydrate in sealed containers at 100°C for 6 min, after having allowed all traces of ether and ethyl acetate to evaporate. The newly-formed diacetyl can be analysed by the method described above. Conversion of acetoin to diacetyl is quantitative, when this technique is employed. Quantitation is possible by preparing calibration curves using known acetoin concentrations which are oxidized and analysed at the same time as "unknowns". New glassware should be used to prevent losses by evaporation made possible by scratches in the glass, where the container is sealed.

C. Analysis of basic fermentation end-products

The amines putrescine, spermine and spermidine are found as structural materials in bacterial cells, where they help to stabilize structures containing acid components. Other amines are found, however, in fermentation supernatants of micro-organisms and generally arise by decarboxylation of α-amino-acids. Little appears to be known about the proportion of any amino-acid decarboxylated by, say, a clostridium rather than deaminated by the Stickland reaction or in other reactions. Accordingly only the qualitative analysis of fermentation amines is carried out. Quantitative analysis awaits further background studies.

Much of the literature on GLC analysis of amines is devoted to the analysis of amines of pharmacological interest, but various American workers have published a number of methods for the analysis of amines of

TABLE VIII

Analysis of basic end-products of fermentation

Author	Growth conditions	Treatment	GLC column	Temperature/detector	Cpds separated
Brooks & Moore, 1969	Chopped meat-glucose for 1 week	CHCl₃ extraction after addition of NaOH to pH 10–11 trifluoroacetylation	3% SE30	80°C for 1 min; C 7·5°C/min to 265°C/dual F.I.D.	Amines
Lambert & Moss, 1973	Peptone-yeast extract with amino-acid 2–4 h	CHCl₃ extraction after addition of NaOH to pH 10 Heptafluorobutyrylation	3% OV-1; 15% Dexsil 300 GC	150°C; 6°C/min to 210°C/F.I.D.	Putrescine, cadaverine
				170°C, 6°C/min to 230°C	
Moss, Lambert & Cherry, 1972	Triple sugar iron agar. Growth into arginine solution for 2·5 h	Dried; N-heptafluoro-butyryl-n-propylesters formed	3% OV-1	150°C for 5 min; 5°C/min to 255°C/ F.I.D.	Citrulline, ornithine, agmatine, putrescine

TABLE IX

Some amines produced by micro-organisms, and their related amino-acids

Name	Formula	Related common amino acids
Methylamine (aminomethane)	CH_3NH_2	Glycine
Ethylamine(aminoethane)	$C_2H_5NH_2$	Alanine
Trimethylamine	$(CH_3)_3N$	—
n-Propylamine(1-aminopropane)	$C_3H_7NH_2$	α-Amino-butyric acid
iso-Butylamine	$(CH_3)_2CH.CH_2.NH_2$	L-Valine
1-Amino-2-propanol	$NH_2.CH_2.CH(OH).CH_3$	L-Threonine
iso-Amylamine	$(CH_3)_2.CH.CH_2.CH_2.NH_2$	L-Leucine
Heptylamine	$C_7H_{15}.NH_2$	—
Pyrrolidine(tetrahydropyrrole)		—
Di-n-butylamine	$(C_4H_9)_2.NH$	—
1,3-Diaminopropane(trimethylene diamine)	$NH_2.CH_2.CH_2.CH_2.NH_2$	—
β-Phenylethylamine(ω-phenylethylamine)	$CH_2.CH_2.NH_2$	L-Phenylalanine
Agmatine	$NH_2(CH_2)_4NH.C(=NH)NH_2$	L-Arginine
Spermine	$NH_2(CH_2)_3.NH.(CH_2)_4.NH.(CH_2)_3.NH_2$	—
Spermidine	$NH_2(CH_2)_4.NH.(CH_2)_3NH_2$	—
Putrescine(tetramethylenediamine)	$NH_2(CH_2)_4.NH_2$	L-Ornithine
Cadaverine(pentamethylenediamine)	$NH_2(CH_2)_5.NH_2$	—
Tryptamine(3-(2-aminoethyl)-indole; ω-aminoethyl(indole)	$CH_2.CH_2.NH_2$	L-Tryptophan

microbiological interest (Table VIII). Table IX lists some of the more important such amines.

The method recommended by Brooks and Moore (1969) is essentially as follows:

After incubation for 1 week at a suitable temperature, 8N sodium hydroxide is added to 6 ml of culture to give a pH 10–11, and amines removed by extraction with 0·5 vol analytical grade chloroform, centrifuging if necessary to break any emulsion which forms upon shaking. The chloroform extract is concentrated by evaporation to 0·09 ml with a stream of air and then trifluoroacetylated. Acylation is performed by addition of about 8·4 μl of trifluoroacetic anhydride (TFA) in pyridine (14:1 by vol). After reaction at 20°C/3 min, 0·07 ml 4% w/v hydrochloric acid is added to decompose excess anhydride and increase ionization of pyridine resulting in its extraction into the aqueous phase which reduces interference with GLC analysis.

Having been allowed to stand for 30 min, the chloroform extract is analysed by GLC on columns of 3% SE-30 on Chromosorb W (80–100 mesh), acid washed, dimethyldichlorosilane treated, high performance support packed in dual 24 ft columns of 0·15 in. i.d. The temperature is programmed at 80°C for an initial period of 1 min, followed by 7·5°C/min to 255°C. The acylation reagent is best freshly prepared from reagents stored under dry conditions. Figure 10 shows a chromatogram of a separation of trifluoroacetyl amines.

If it is desired to analyse putrescine, citrulline, ornithine or agmatine, the method of Moss et al. (1972) may be employed. This provides more precise information on an organism's arginine dihydrolase system than would detection of ammonia alone.

A dense inoculum of growth from a triple sugar–iron slope is inoculated into 0·3 ml of 2·5 mM L-arginine at pH 6·8. Incubation at 37°C for 2·5 h results in breakdown of arginine if the relevant enzymes are present in the test organism.

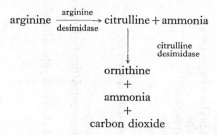

Some organisms use different reactions in which arginine is converted into urea and ornithine by arginase; or into agmatine by arginine decarboxylase.

The supernatant is analysed after drying at 100°C and forming N-hepta-fluorobutyryl-n-propyl ester derivates. Propylation is brought about in 10 min at 100°C with 8N hydrogen chloride in propan-1-ol. The propyl esters are evaporated to dryness in a stream of dry nitrogen at 100°C. 0·2 ml heptafluorobutyryl anhydride (Pierce Chemical Co.) and 0·1 ml ethyl acetate are added and after sealing the tube acylation proceeds at 150°C/

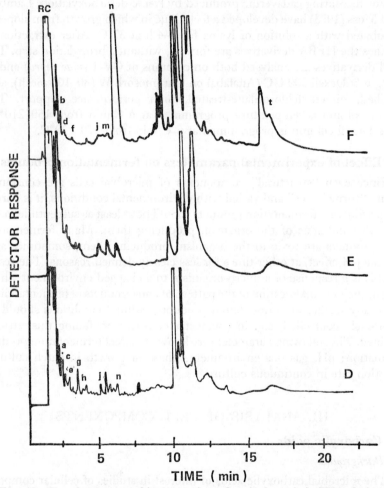

FIG. 10. Gas chromatographic separation of amines as their TFA derivatives (after Brooks and Moore, 1969). Key: (a) methylamine, (b) trimethylamine, (c) ethylamine, (d) n-propylamine, (e) iso-butylamine, (f) 1-amino-propan-2-ol, (g) iso-amylamine, (h) pyrrolidine, (j) di-n-butylamine, (m) 1,3-diaminopropane, (n) β-phenylethylamine, (t) spermine. Reproduced by kind permission of the National Research Council of Canada.

10 min. At room temperature samples are evaporated under dry nitrogen, taken up in 0·1 ml ethyl acetate and analysed. GLC analysis is on 12 ft by 0·25 in. o.d. glass columns of 3% w/v OV-1 on Chromosorb W (80–100 mesh) acid-washed, dimethylchlorosilane-treated high performance support. An isothermal period of 150°C/5 min is followed by temperature programming at 5°C/min to 255°C.

For measuring cadaverine produced by lysine decarboxylation, Lambert and Moss (1973) have developed a technique in which growth from slopes is incubated with a solution of lysine for 2–4 h at 37°C. After extraction of amines the HFBA derivatives are formed, without a propylation step. The acyl derivatives are analysed both on columns of OV-1 (*vide supra*) and of 15% w/v Dexsil 300 GC (Analabs) on Chromosorb W (80–100 mesh), acid washed, dimethylchlorosilane-treated, high performance support. The OV-1 column is temperature programmed at 6°C/min from 150–210°C. The Dexsil column is programmed at 6°C/min from 170–230°C.

D. Effect of experimental parameters on fermentation products

Since even "structural" components of microbial cells are constantly being "turned over" and varied with environmental conditions, it is hardly surprising that fermentation products should be at least as susceptible to the growth conditions of the organism producing them. Many fermentation end-products are toxic to the organism producing them, sometimes as a simple pH effect, at other times for less clearly defined reasons. The result of this is a response of a micro-organism to a changed environment which might include modification of the pattern of end-products of fermentation.

In any studies on fermentation products, cultural conditions should be rigorously controlled, e.g. in chemostat cultures; or, failing this, strictly defined. The following parameters are known to affect fermentation product formation; pH, gaseous environment, phase of growth in batch culture; dilution rate in continuous culture.

III. ANALYSIS OF CELL COMPONENTS

A. Carboxylic acids

1. *Background*

The microbial carboxylic acids of interest in studies of cellular composition are of longer chain length (12 up to 22 carbon atoms) than the fatty acids described above as fermentation end-products. Much of the longer chain acid is found esterified, in simple and complex lipid, in cell membranes and in the cell envelope of Gram-negative organisms. Shorter chain intracellular fatty acids are to be found as intermediates in biosynthesis of

III. GAS–LIQUID CHROMATOGRAPHIC CHEMOTAXONOMY

TABLE X

Some intracellular bacterial fatty acids

Homologous series	Name of acid	Formula
Saturated straight chain	Dodecanoic (lauric)	$CH_3(CH_2)_{10}.COOH$
	Tridecanoic (tridecylic)	$CH_3(CH_2)_{11}.COOH$
	Tetradecanoic (myristic)	$CH_3(CH_2)_{12}.COOH$
	Pentadecanoic (pentadecylic)	$CH_3(CH_2)_{13}.COOH$
	Hexadecanoic (palmitic)	$CH_3(CH_2)_{14}.COOH$
	Heptadecanoic (margaric)	$CH_3(CH_2)_{15}.COOH$
	Octadecanoic (stearic)	$CH_3(CH_2)_{16}.COOH$
	Nonadecanoic (nondecylic)	$CH_3(CH_2)_{17}.COOH$
	Eicosanoic (arachidic)	$CH_3(CH_2)_{18}.COOH$
	Heneicosanoic (—)	$CH_3(CH_2)_{19}.COOH$
	Docosanoic (behenic)	$CH_3(CH_2)_{20}.COOH$
Unsaturated straight chain (*monoenoic*)	Δ^9-tetradecenoic (myristoleic)	$CH_3(CH_2)_3.CH=CH(CH_2)_7.COOH$
	Δ^9-hexadecenoic (palmitoleic)	$CH_3(CH_2)_5.CH=CH(CH_2)_7.COOH$
	cis-Δ^9-octadecenoic (oleic)	$CH.(CH_2)_7.CH_3$ ∥ $CH.(CH_2)_7.COOH$
	cis-Δ^{11}-octadecenoic (*cis*-vaccenic)	$CH.(CH_2)_5.CH_3$ ∥ $CH.(CH_2)_9.COOH$
	trans-Δ^9-octadecenoic (elaidic)	$CH_3(CH_2)_7.CH$ ∥ $CH.(CH_2)_7.COOH$
	Δ^9-eicosenoic (gadoleic)	$CH_3(CH_2)_9.CH=CH(CH_2)_7.COOH$
(*dienoic*)	*cis-cis*-$\Delta^{9,12}$-octadecenoic-(linoleic)	$CH_3(CH_2)_4.CH$ ∥ $CH.CH_2CH$ ∥ $CH.(CH_2)_7.COOH$
(*trienoic*)	$\Delta^{9,12,15}$-octadecenoic-(linolenic)	$CH_3CH_2CH=CH.CH_2.CH$ ∥ $COOH(CH_2)_7.CH.CH_2.CH$
Branched chains (*iso*)	14-methyl-pentadecanoic(*iso*-palmitic)	$(CH_3)_2CH(CH_2)_{12}.COOH$
(*anteiso*)	13-methyl-pentadecanoic-(*anteiso*-palmitic)	$CH_3.CH_2.CH(CH_3).(CH_2)_{11}.COOH$
Cyclopropane	*cis*-11,12-methylene octadecanoic-(lacto-bacillic)	$CH_3.(CH_2)_5.CH—CH.(CH_2)_9.COOH$ \\/ CH_2

the longer chain acids. It appears that three general classes of micro-organisms exist with respect to percentage lipid composition:

(i) Gram-positive organisms with $\approx 3\%$ lipid.
(ii) Gram-negative organisms with $\approx 11\%$ lipid.
(iii) *Microbacterium, Nocardia* with $>20\%$ lipid, which includes extra-long chain hydroxy fatty acids of up to 88 carbon atoms.

Carboxylic acids found intra-cellularly are comprised of many different classes of fatty acid, viz., saturated fatty acids either normal, iso, or ante iso-isomers, unsaturated acids, mono, di-, or tri-, enoic acids with various positions of double bond, hydroxy fatty acids, and cyclo-propane fatty acids. Some of the commoner isomers found are shown in Table X. A standard chromatogram is shown in Fig. 11.

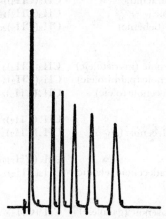

FIG. 11. Gas chromatogram of carboxylic methyl ester standards on PEGA at 190°C; peaks in order of elution-methyl esters of C_{11}, C_{12}, C_{13}, C_{14}, C_{15} straight chain saturated fatty acids.

Methods for the analysis of these acids have to take into account the following facts:

(i) Cellular fatty acid composition is dependent, in part, upon growth conditions, as well as genotype of the organism.
(ii) The carboxylic acids are mainly present not as free acids but covalently linked to other molecules.
(iii) Free fatty acids of 12 carbon atoms and above are not sufficiently volatile for satisfactory analysis by GLC as free acids.
(iv) Final analysis will to some extent depend upon extraction procedures employed.

2. Cultural conditions

The effect of cultural conditions upon fatty acid profiles of bacteria was shown by Marr and Ingraham (1962) who used a chemostat to control the environmental conditions accurately in their work. They showed that fatty acid composition could be modified by change in temperature.

More recently Farshtchi and McClung (1970) have shown that substrate can affect the lipid composition of *Nocardia asteroides*. The effects of numerous experimental parameters have now been studied including oxygenation (Drucker and Owen, 1973), pH (Drucker, Griffith and Melville, 1973), dilution rate and temperature in continuous culture (Marr and Ingraham, 1962) and vitamin and magnesium limitation (Drucker, Griffith and Melville, 1974b). All the above parameters have been shown to have an effect upon fatty acid profiles in the case of streptococci grown in a chemically defined medium. Ideally organisms would be grown under carefully controlled conditions in a chemostat, but this would be too demanding a method for routine use. Therefore a compromise is usually employed, a precise inoculum of 18 h culture of the test organisms is grown in a rich medium under standard conditions of oxygenation, at a fixed and accurately controlled temperature, with constant shaking for a precise length of time, and then harvested under standard conditions. The author has employed the following technique in the analysis of streptococci. Each test strain is checked for purity by growth on 5% horse-blood agar, Gram-staining and serological grouping and is then heavily inoculated into triplicate 2 ml aliquots of "Oxoid" brain–heart infusion in bijou bottles and grown for 18 h at 37°C. After growth, starter cultures are added to a further 60 ml "Oxoid" brain–heart infusion in 100 ml "medical flats". These are placed in an anaerobic jar with the caps loose, and the air over the culture is replaced by an atmosphere of 5% CO_2/95% air. The caps are then rapidly screwed down and the bottle is incubated in a shaking water bath at 37°C for 48 h. Shaking is employed not for aeration but to discourage stagnation caused by heavy "fall-out" of long chains of organisms. Residual oxygen in the culture is soon removed during growth. After growth, cultures are re-checked for purity and harvested by centrifugation at 3000 g/20 min at 4°C, then washed with Sorensen's buffer, once, and distilled water twice, and finally freeze-dried (without serum) in an "Edwards" freeze-dryer, before extraction of carboxylic acids. Some of the different cultural conditions which various authors have satisfactorily employed are shown in Table XI.

3. Extraction of intracellular carboxylic acids and derivative formation

Methods for removing carboxylic acids from microbial cells and convert-

TABLE XI

Analysis of intracellular fatty acids of micro-organisms

Author ref.	Preparation of carboxylic esters		Detector	Temperature	GLC analysis		Chain length studied
	Saponification conditions	Methylating reagent			Polar columns	Non-polar column	
(1) Agate & Vishniac, 1973	15% KOH/50% CH$_3$OH	BF$_3$/CH$_3$OH	F.I.D.	120–150°C (polar) 170–230°C (non-polar)	18% DEGS	3% OV1	C$_8$–C$_{20}$
(2) Drucker, 1974	—	BF$_3$/CH$_3$OH	F.I.D.	190°C	10% PEGA	3% OV1	C$_{14}$–C$_{22}$
(3) Ifkovits & Ragheb, 1968	5% KOH/95% C$_2$H$_5$OH	CH$_3$N=NCH$_3$	F.I.D.	170°C (polar) 190°C (non-polar)	15% DEGA	15% Apiezon L	C$_{11}$–C$_{20}$
(4) Kaneda, 1968	0·5 g KOH/50 ml CH$_3$OH/ 25 ml cell suspension	CH$_3$N=NCH$_3$	F.I.D.	170°C (polar) 130°C–>260°C (non-polar)	7% PEGA	2·5% SE-30	C$_{12}$–C$_{17}$
(5) Kimble et al., 1969	—	BF$_3$/CH$_3$OH	F.I.D.	170°C, 180 or 190°C (polar) 160°C–>250°C (non-polar)	15%† EGS	3·8%† SE-30	C$_8$–C$_{21}$
(6) Moss & Lewis, 1967	15% KOH/50% CH$_3$OH	BF$_3$/CH$_3$OH	F.I.D.	110–>195°C	12% EGA	2% SE-30	C$_{10}$–C$_{20}$
(7) Thoen et al., 1971	3 ml 33% KOH/15 ml CH$_3$OH	CH$_3$N=NCH$_3$	F.I.D.	200°C (polar) 250°C (non-polar)	15% DEGS	?% SE30	C$_8$–C$_{24}$

| (8) Wade & Mandle, 1974 | $(CH_3)_4N^+OH^-$ in CH_3OH | — | — | F.I.D. 80->195°C | — | — | 10% DEGS | — | — | C_{10}–C_{21} |
| (9) White & Frerman, 1968 | — | — | — | F.I.D. 156°C | — | — | 15% EGA | — | — | — |

† Other stationary phases used including Apiezon L, neopentyl glycol succinate, carbowax 20M, FFAP, and benzylamine adipate.

| Ref. | Additional analytical techniques | | | | | Data analysis | |
	Mass-spectrometry	TLC	Bromination	Hydrogenation	I.R.	Integration	Computer
(1)	—	—	—	—	—	Disc-integration	—
(2)	—	—	—	—	—	Electronic digital	Computation of ECN % peak areas and statistical analysis
(3)	—	Purification of methyl ester fraction	+	—	—	—	—
(4)	+	—	+	+	+	—	—
(5)	—	Separation of mercuric acetate adducts on silicic acid impregnated with $AgNO_3$	—	+	+	—	—
(6)	—	—	—	—	—	Disc-integration	—
(7)	—	—	—	—	—	Disc-integration	—
(8)	—	—	—	—	—	—	—
(9)	—	Mercuric adducts of monoeoate concentrated on silicic acid	—	+	—	—	—

ing them into volatile derivatives suitable for GLC are legion, and are summarized in Table XI. Basically two approaches are employed:

(i) removal of intracellular fatty acids under alkaline conditions (saponi-fication) followed by conversion of carboxylic sodium, or potassium, salts to volatile esters.

(ii) removal of intracellular fatty acids under acidic conditions coupled with conversion to volatile carboxylic esters (trans-esterification).

The former approach results in more efficient removal of fatty acids, especially hydroxy-acids, but the latter approach is generally less time-consuming.

(i) *Saponification.* Conditions for saponification are not uniform and include refluxing with potassium hydroxide/methanol/water (3 : 90 : 10, w/v/v) for 30 min (Brian and Gardner, 1967); heating for 1 h at 100°C in 5% sodium hydroxide in 50% aqueous methanol (Moss, Lambert and Merwin, 1974); sonic disruption in 80% v/v aqueous ethanol followed by saponification in 95% aqueous ethanol containing 5% w/v potassium hydroxide at room temperature, under nitrogen (Ifkovits and Ragheb, 1968).

After saponification, carboxylic esters are generally recovered prior to esterification by acidification and extraction into a suitable lipid solvent. Again, numerous techniques have been suggested; usually, the saponified samples are adjusted to pH 2·0 with sulphuric acid of varying strengths, e.g. 15% w/v or 50% w/v (Thoen *et al.*, 1971; Brian and Gardner, 1967) or, 6N hydrochloric acid after prior removal of non-saponifiable lipid using 20% v/v chloroform-hexane (Moss *et al.*, 1974). Extraction techniques by the workers mentioned above have used various solvents, such as chloro-form-hexane, petroleum ether, and hexane, either specially purified or not, with or without a water wash, and with drying over anhydrous sodium sulphate or not according to the preference of the laboratory concerned. The free acids are then esterified by a variety of procedure, usually resulting in production of carboxylic methyl esters which are readily separated by GLC. Methylation is possible by a number of techniques based on the use of (a) methanol-mineral acid, (b) boron trihalide-methanol, (c) diazo-methane, (d) quaternary ammonium compounds.

(a) *Methanol-mineral acid:* usually, anhydrous methanol containing anhydrous hydrogen chloride (Fischer's Reagent) is used by Metcalfe and Schmitz (1961) or by Vorbeck *et al.* (1961) whose method, based on that of Stoffel *et al.* (1959) methylates free carboxylic acids, after solution in benzene, with 5% w/v hydrogen chloride in super dry methanol. After refluxing, water is added and the methyl esters are taken up in petroleum spirit and dried over a mixture of anhydrous sodium sulphate and sodium bicarbonate. Hornstein *et al.* (1960) describe an unusual procedure in which acids are absorbed on to Amberlite IRA-400 and esterified *in situ* with Fischer's reagent. Methyl carboxylic esters are then eluted with petroleum ether and dried over anhydrous sodium sulphate. Hydrogen chloride may be replaced by the 1% v/v sulphuric acid in methanol, which is refluxed for 30 min with the acids to be methylated, followed by addition of 0·6 vol water and 2·0 vol chloroform into which the esters are extracted.

(b) *boron trihalide-methanol:* two reagents have been used—boron trifluoride or boron trichloride. These react with carboxylic acids (or salts of the acids) in the presence of methanol to yield a carboxylic methyl ester, hydrogen halide (or halide salts) and boric acid. Such reactions take place very rapidly and virtually quantitatively, at moderate temperatures. The procedure advocated by Metcalfe and Schmitz (1961) for preparation of methyl esters of carboxylic acids with 11 or more carbon atoms is as follows:

One hundred to 200 mg of fatty acid is placed in a 20 mm × 150 mm test tube and 3 ml reagent is added. After boiling for 2 min and allowing to cool, the reaction mixture is shaken with 30 ml of petroleum ether and 20 ml distilled water. The petroleum ether phase is filtered into a 50 ml beaker and the solvent removed by evaporation on a 60°C waterbath. Brian and Gardner (1967) favoured the use of 5 ml of a 10% w/v solution of boron trichloride in methanol, under the same conditions as employed by Metcalfe and Schmitz (1961).

(c) *diazomethane:* this has been widely used but has two disadvantages; it has to be freshly prepared and it is liable to explode if incorrectly handled. The method most widely used is probably that of Schlenk and Gellerman (1960), whose technique for the esterification of fatty acids with diazomethane is as follows: After passing through ether in tube 1, nitrogen bubbles through a mixture, in tube 2, of 0·7 ml 2-(β-ethoxyethoxy)-ethanol (carbitol), 0·7 ml diethyl ether, and 1 ml of potassium hydroxide solution (6 g KOH: 10 ml H_2O) to which is added 2 mmoles of N-methyl-N-nitroso-p-toluenesulphonamide per milliequivalent of fatty acid to be esterified. This mixture generates the yellow gas diazomethane and as soon as a yellow colour appears, connection is made with tube 1. Diazomethane is now carried into tube 3. Tube 3 contains a few mg fatty acids in 2 ml of ether–methanol (9 : 1). Methylation is complete when excess diazomethane appears in tube 3. Removal of this excess is possible in a stream of nitrogen.

(d) *quaternary ammonium compounds:* a technique related to the traditional use of alkali to saponify lipids is the use of quaternary alkyl ammonium hydroxides which form thermolabile salts of carboxylic acids capable of being decomposed in the heated injection part of the gas chromatograph, to liberate carboxylic alkyl esters. Usually the reagent used is tetramethyl-ammonium hydroxide which results in formation of carboxylic methyl esters (Wade and Mandle, 1974) and is favoured by some as a "one-step preparation reagent", although the technique has been criticized by other workers on the grounds that artefactual peaks may be produced upon GLC.

(ii) *Transesterification.* Techniques for transesterification are essentially the same as those described above for esterification of saponified lipid. The method employed routinely by the author is based upon the use of boron trifluoride-methanol, which is now commercially available as a 15% solution, used in the following way:

Approximately 10 mg aliquots of freeze-dried bacterial cells are weighed into freeze drying ampoules which are carefully restricted using an "Edwards" ampoule restrictor. After cooling, 1·0 ml of boron trifluoride-methanol "methylating reagent" is added, the ampoule is evacuated and sealed. Transesterification is carried out in a boiling water bath for 1 h. After cooling, the contents of the ampoule are added to 9·0 ml water in a universal bottle and shaken with 1 vol analytical grade petroleum spirit or heptane. Failure to shake continuously results in a thick emulsion forming which is difficult to break, even with centrifugation. Some emulsion forms, with bacteria having high lipid contents, even if these precautions are observed and the emulsion can then be broken either by giving the universal bottle a sharp tap, or by centrifuging at 2000 *g* for 20 min.

The aqueous phase is re-extracted with petroleum spirit and the organic phases are pooled, and stored at $-18°C$ until required for analysis. The technique which is very simple in use requires very clean (preferably unused) glassware in the form of universal bottles with metal caps without liners. Rubber liners or plastic tops result in artefacts appearing upon GLC analysis. Extraction in universal bottles is preferable to using separating funnels since tap grease dissolves in the petroleum spirit used as an extraction solvent.

When a sample is required for analysis, excess solvent is removed by evaporation, in a gentle stream of air in a well-ventilated fume cupboard, and the sample concentrated down to a volume of 20 μl. Brian and Gardner (1967) and Metcalfe and Schmitz (1961) believe boron halides to produce analyses similar to those obtained using less modern techniques. Moss *et al.* (1974) found boron halides to yield artefacts, although there is no indication that air was deliberately excluded from the reaction vessel in their experiments. Vorbeck *et al.* (1961) have compared various methylation techniques.

4. *Gas–liquid chromatographic analysis of carboxylic alkyl esters*

The discussion which follows will be restricted to the methods employed in the analysis of methyl esters of carboxylic acids derived from bacterial lipids. Simple GLC analysis can be backed up with derivative formation and separation of lipid classes by thin layer chromatography (TLC), by techniques such as, bromination of enoates, or hydrogenation, or by mass-spectrometry either of on the "fly"—or "trapped" samples. If the researcher is interested simply in a chemotaxonomic GLC fingerprint, the absolute identity of peaks is not of paramount importance, and the related procedures allied to GLC are not generally employed.

(i) *Conditions for GLC analysis.* The GLC analysis of methyl esters of longer chain fatty acids is well established and a summary of some of the analytical parameters employed appears in Table XI. Usually a polar column is used for the analysis, either diethylene glycol succinate (DEGS), diethylene glycol adipate (DEGA) or polyethylene glycol adipate (PEGA), although analysis on a non polar silicone column (OV1 or SE30) is frequently carried out to provide corroborative evidence of identifications made tentatively on the basis of analyses on polar columns. Figure 12 shows a standard mixture analysed on polar and non-polar columns. Polar columns tend to separate esters of saturated acids from enoates of the same chain length; whereas on non-polar columns the retention volume is determined mainly by chain length of the fatty acid rather than by degree of

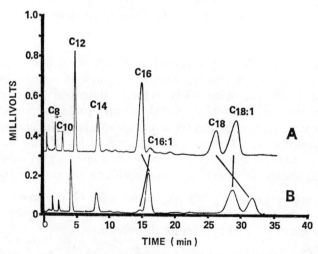

Fig. 12. The effect of stationary phase polarity on the separation of carboxylic methyl esters by GLC. Key: (A) Reoplex, (B) Apiezon L (after Orr and Callen, 1958). Reproduced by kind permission of the American Chemical Society.

unsaturation. The method employed in the author's laboratory is as follows:

Petroleum spirit solutions of esters, after evaporation down to 20 μl, are injected (2 μl) on to a 5 ft by $\frac{1}{4}$ in. o.d. glass column of 10% PEGA on acid-washed diatomite C run at 190°C isothermally in a PYE 104 gas chromatograph, equipped with gas flow controllers and a flame ionization detector, using a nitrogen carrier gas flow of 40 ml/min. If an electronic integrator is to be used to measure peak areas then a wide range amplifier, linear over at least five decades, must be used to prevent large peaks causing "saturation" of the amplifier and causing the integrator to print out two or more retention times for a single peak. A typical GLC analysis is shown in Fig. 6. A typical integrator print out is shown in Fig. 13. Corroboration of peak

01	16504
24	2997997
51	5569
113	63343
148	77169
168	230
195	84589
259	100749
349	122161
559	2604
783	226
1086	485
1175	9
1176	60

FIG. 13. Integrator print out, showing retention times in seconds and peak areas.

identities can be provided by gas chromatography of standards analysed on their own; or, preferably, admixed with the sample (co-chromatography), in addition to the use of non-polar columns described above. A typical chromatogram appears in Fig. 14. "Unknown" peaks can sometimes be tentatively identified in the absence of the appropriate standard by plotting a graph showing log RT vs. no. carbon atoms (Fig. 15) for the various homologous series of saturated and unsaturated fatty acids (Table X), and from which retention times of acids not available can be accurately predicted.

(ii) *Chemical techniques used in conjunction with GLC.* Since fatty acid mixtures derived from bacteria may be immensely complex, GLC is frequently supplemented by additional procedures.

(a) *TLC–GLC:* thin layer chromatography can be used to separate homologous series of enoates prior to GLC as follows: Methyl carboxylic esters, after removal of virtually all the petroleum spirit solvent by evaporation in a

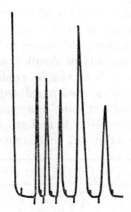

FIG. 14. Gas chromatogram of carboxylic methyl esters, with methyl palmitate, note increased peak area of methyl palmitate (cf. Fig. 11).

Thunberg tube, are treated with 1·0 ml of Mangold's reagent (Stahl, 1969) containing 14 g mercuric acetate, 2·5 ml H_2O, 1 ml acetic acid, in 250 ml methanol, and the tube is evacuated to remove oxygen. Reaction takes place

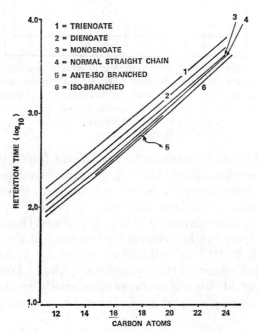

FIG. 15. Relationship between carbon number and $\log_{10}$ (retention time), for homologous series of carboxylic esters.

under nitrogen, in the dark, at room temperature over a period of 48 h, to produce the esters' acetoxymercurimethoxy-derivatives. Excess methanol is removed *in vacuo* and the residue is dissolved in 1 ml chloroform, which is washed twice with 1 ml aliquots of distilled water in order to remove mercuric reagent. The derivatives can be readily separated into enoate classes by TLC after spotting chloroform extracts on plates of Kieselgel G, which are run unidirectionally at room temperature in two solvents, viz., petroleum spirit–diethyl ether (80 : 20) and propan-1-ol–glacial acetic acid (100 : 1); the latter solvent is run for 2 cm less than the former (Stahl, 1969).

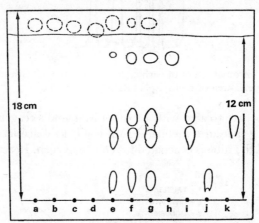

FIG. 16. The separation of acetoxymercurimethoxy derivatives of carboxylic methyl esters into enoate classes by thin layer chromatography (after Stahl, 1969). Key: (a) stearate, (b) oleate, (c) linoleate, (d) linolenate, (e–j) derivatives of (e) C_{16} esters of *Chlorella*, (f) C_{18} esters of *Chlorella*, (g) all methyl esters of *Chlorella*, (h) oleate, (i) linoleate, (j) linolenate, (k) mercuric acetate. Reproduced by kind permission of Springer-Verlag, Berlin.

The saturated esters lie between the two solvent fronts; the unsaturated acid esters having increasingly lower R_f values as their degree of unsaturation increases lying below the second solvent front. Visualization of the enoate acetoxymercurimethoxy derivatives is possible by spraying with 0·1% w/v, S-diphenyl-carbazone in 96% v/v ethanol (Fig. 16).

Unsprayed spots can be recovered by removal of the thin layer and treatment with N HCl in methanol to re-generate enoate followed by extraction with 2 ml diethyl ether–petroleum spirit (1 : 1) to recover esters for GLC. Separated classes of esters can be separately gas chromatographed and yield simpler chromatograms, more amenable to accurate interpretation (Fig. 17).

Other techniques for simplifying GLC analysis include bromination and hydrogenation.

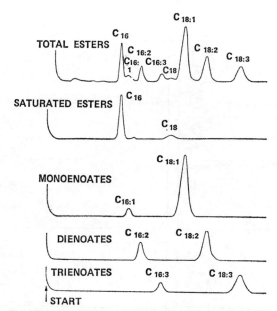

FIG. 17. Chromatograms of carboxylic esters after separation into enoate classes, by thin layer chromography of acetoxymercurimethoxy derivatives, followed by recovery of esters and GLC analysis (after Mangold and Kammereck, 1961). Reproduced by kind permission of the authors.

Bromination to produce addition across the double bonds of enoates results in formation of methyl esters of bromo-carboxylic acids whose decreased volatility results in removal of enoate from the gas chromatogram.

Hydrogenation converts enoic acid esters to their corresponding saturated esters, resulting in disappearance of peaks due to esters of unsaturated acids, and corresponding enlargement of peaks due to esters of saturated acids.

(b) *IR–GLC and MS–GLC:* additional instrumentation used in conjunction with GLC includes infrared spectroscopy and mass spectrometry. Brief descriptions of applications of these techniques are as follows:

(i) Infrared spectroscopy was found useful as an adjunct to GLC by Kimble *et al.* (1969) who employed the technique on various methyl ester samples and lipid fractions of *Clostridium botulinum* on alkali-halide, before, and after, mild hydrogenation; they confirmed the presence of cyclopropane-carboxylic acid by the disappearance of the infrared band at 1020 cm^{-1} and its replacement by a band at 1370 cm^{-1} which is the wavenumber at which methyl branched alkanes characteristically absorb.

(ii) *GLC–MS.* The GLC provides separation but not identification of

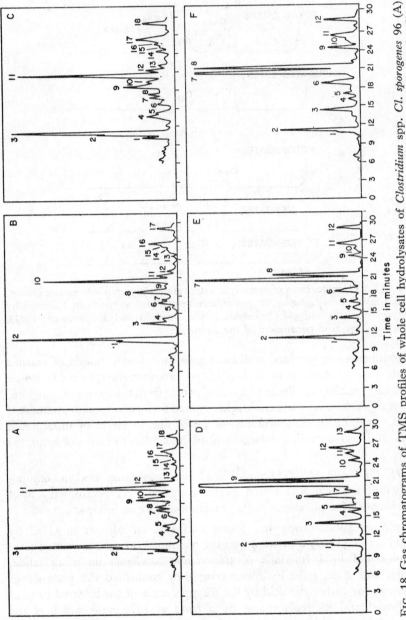

Fig. 18. Gas chromatograms of TMS profiles of whole cell hydrolysates of *Clostridium* spp. *Cl. sporogenes* 96 (A); *Cl. sporogenes* 81 (B); *Cl. sporogenes* 1783 (C); *Cl. botulinum* 16 type A (D); *Cl. botulinum* 17 type B (E); *Cl. botulinum* 28 type A (F); (continued on p. 97).

(continued on p. 97).

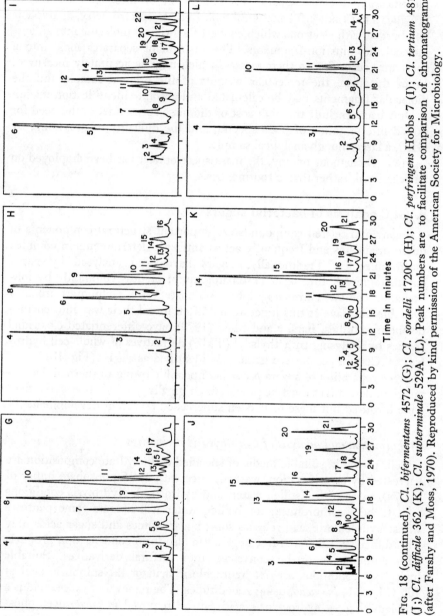

FIG. 18 (continued). *Cl. bifermentans* 4572 (G); *Cl. sordelli* 1720C (H); *Cl. perfringens* Hobbs 7 (I); *Cl. tertium* 481 (J;) *Cl. difficile* 362 (K); *Cl. subterminale* 529A (L). Peak numbers are to facilitate comparison of chromatograms (after Farshy and Moss, 1970). Reproduced by kind permission of the American Society for Microbiology.

mixtures of volatile compounds. The mass spectrometer (MS) provides identification but not separation. The union of these two techniques provides a most powerful analytical tool. In mass-spectrometry, a sample is bombarded with electrons which convert the original molecule into an ion-molecule or into ion-fragments. The instrument separates ions, into a spectrum, according to their masses, which may be accurately measured. From this data, the molecular weights of the original sample and the molecular fragments may be calculated and absolute identification made.

Drawbacks include the high cost of the mass spectrometer, the need for skilled interpretation of mass spectra, and the difficulty of providing, by GLC, a large enough analytical sample.

In GLC-chemotaxonomy, the mass-spectrometer has been employed on an occasional, rather than a routine, basis.

B. GLC analysis of bacterial sugars

Sugars and related compounds are important structural components of micro-organisms and frequently act as antigenic determinants in complex macromolecules. Traditionally, sugars have been analysed by paper chromatography, thin layer chromatography; or, less commonly by ion-exchange column chromatography. Adequate GLC techniques are now available but have at this juncture not often found their way into chemotaxonomic studies. Farshy and Moss (1970) have differentiated clostridial species by analysing trimethylsilyl (TMS) derivatives of whole cell hydrolysates. Certain peaks were tentatively identified as sugars (Fig. 18).

Since separation of sugars *per se* is impossible owing to thermal decomposition, derivatives must be prepared. Originally, methyl ether derivatives were favoured but these have been superceded by recent developments.

(i) *Formation and analysis of trimethylsilyl derivatives*

Cells grown in suitable media of specified carbohydrate composition are harvested by centrifugation, washed, once with M/15 phosphate buffer of pH 7·0, twice with distilled water, and hydrolysed. Hydrolysis conditions tend to be a compromise; for boiling water will liberate some pentoses whilst hydrochloric acid releases some sugar amines and sugar acids only with difficulty. Liberated free sugars can also form condensation products with amino-acids or be converted into furfural derivatives. Suitable hydrolysis conditions are 2N hydrochloric acid or 2N sulphuric acid at 100°C for 2 h, *in vacuo*, using sealed tubes. The removal of excess acid is a problem. Excess hydrochloric acid can be removed *in vacuo* over phosphorous pentoxide and potassium hydroxide pellets (separate containers!) but evaporation of hydrogen chloride from samples takes place only after a preliminary concentration of acid which may be undesirable. Excess sulphuric acid cannot be removed by evaporation and is usually neutralized with barium hydroxide, which converts the acid into insoluble barium sulphate.

If neutral sugars only are required, the hydrolysates can be filtered through anionic- and cationic-, exchange resins, which remove many interfering substances, then re-dried.

The dried neutral sugars can be silylated in various ways including the following.

A maximum of 10 mg of dried sugars in a stoppered vial is dissolved in 1·0 ml of a mixture of anhydrous pyridine (dried by storage over potassium hydroxide pellets), 0·2 ml of hexamethyl disilazane, and 0·1 ml trimethylchlorosilane. The mixture is vigorously shaken for 30 min and the reaction allowed to proceed to completion by warming to 60°C for several hours. Cloudiness caused by insoluble particles of ammonium chloride may be apparent at this stage. The suspension may be centrifuged and the supernatant evaporated just to dryness under nitrogen, then taken up in the minimum amount of heptane and re-concentrated. 1 μl aliquots may be analysed by GLC.

The conditions preferred by Mitruka (1974) require facilities for temperature-programming. Dual glass columns of 0·3 cm i.d. and between 7 and 17 m in length are packed with 3% OV-1 on Chromosorb W (AW-DMCS.H.P.). The column oven is programmed at 5°C/min from 80°C. Other parameters are: injection temperature, 250°C, detector-temperature, 290°C.

In addition to the aqueous acid-extraction technique outlined above, it is possible to heat lyophilized cells with hydrogen chloride in methanol then silylating reagent to produce chromatograms of the type illustrated by Fig. 18. The technique of Farshy and Moss (1970) employs 20 mg freeze-dried cells which are extracted for 8 h at 100°C with 20 ml 1N hydrogen chloride in methanol.

After cooling and drying, the residue is repeatedly taken up in methanol and dried down, until all traces of hydrogen chloride have been removed. The neutral residue is dissolved in 1 ml of potassium hydroxide-dried pyridine, then 0·3 ml hexamethyldisilazane and 0·15 ml trimethylchlorosilane are added. After preliminary shaking, the reaction mixture is allowed to stand overnight at room temperature, and heated as described above. Analysis is on both polar and non-polar columns, 6-ft glass columns of 1% OV-17 on 80–100 mesh high performance Chromasorb G temperature programmed at 5°C/min from 120–200°C and 8 ft glass columns of 3% OV-1 on 60–80 mesh Gas-Chrom Q temperature programmed at 3°C/min from 130–220°C, then raised to 250°C and run isothermally for 10 min.

C. GLC of amino-acids

Amino-acids are one of the basic building blocks in the chemical architecture of cells, but their analysis for taxonomic purposes has so far been restricted almost entirely to cell wall amino-acid composition.

This apparent lack of interest in amino-acids of whole cells is based partly on the fact that production of large quantities of a single inducible enzyme might alter an organism's overall amino-acid composition. Also, until recently, suitable, rapid, GLC methods for amino-acid analysis were

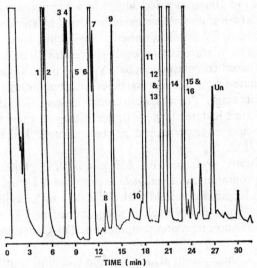

Fig. 19. Analysis of treponemal amino acids by gas chromatography of their N-trifluoroacetyl-n-propyl ester derivatives (after Moss *et al.*, 1971). Key: (1) alanine, (2) glycine, (3) serine, (4) threonine, (5) valine, (6) leucine, (7) *iso*-leucine, (8) cysteine, (9) proline, (10) methionine, (11) aspartic acid, (12) phenylalanine, (13) ornithine, (14) glutamic acid, (15) lysine, (16) tyrosine. Reproduced by kind permission of the British Medical Association.

not available; now a number of techniques are possible for amino-acid analysis. The amino-acid composition of whole cell hydrolysates of treponemes has in fact been reported by Moss *et al.* (1971; Fig. 19).

1. *Derivative formation and analysis*

Freeze dried cells (approx. 10 mg) are hydrolysed with 2 ml 6N hydrochloric acid *in vacuo*, in sealed tubes, at 106°C for 18 h. After cooling, lipids can be removed by extraction with organic solvents. Excess acid is removed by evaporation and neutral compounds are separated on a column of Amberlite CG-120 ion-exchange resin (100–200 mesh, hydrogen form)

(Moss *et al.*, 1971a). Amino-acids are absorbed, and eluted later with 3N ammonium hydroxide. After drying, the non-volatile amino-acids are converted to more volatile derivatives, e.g. N-trifluoroacetyl-*n*-propyl, N-heptafluorobutyryl-*n*-propyl (Moss *et al.*, 1971b), N-trifluoroacetyl-*n*-butyl (Gehrke and Shahrokhi, 1966) or N-trimethylsilyl (Gehrke *et al.*, 1969) derivatives. The problems faced are the relative volatility of glycine's, and alanine's, derivatives and the relative difficulty in forming derivatives of some of the basic amino-acids. In many techniques, resolution of leucine and iso-leucine is very poor. The various experimental conditions which have been used for GLC analysis of amino-acid esters are summarized in Table XII.

An interesting alternative to esterification is reaction of amino-acids with ninhydrin in a quantitative reaction which converts α-amino-acids of (n) carbon atoms to carbon dioxide, ammonia and an aldehyde of $(n-1)$-carbon atoms. Unfortunately, some of the lower aldehydes are extremely volatile, e.g. formaldehyde b.pt. $-21°C$; and if a F.I.D. detector is used formaldehyde (derived from glycine) is not detected (Zlatkis *et al.*, 1960).

D. Effect of experimental parameters on cellular components

The effect of environment on cellular composition is well known. For example, it is customary for streptococci to be grown in glucose broths prior to extraction of the Lancefield carbohydrate antigen.

However, these effects have been studied, from the chemotaxonomic viewpoint, mostly with respect to fatty acid profiles of cells, which show changes with type of culture (whether batch or continuous), age of cells, temperature, pH, carbohydrate energy source, vitamin limitation, and trace metal limitation. Apart from *actual* changes in chemotaxonomic finger-prints it is also possible to detect *apparent* changes resulting merely from change in analytical procedures. The latter changes may result from use of different columns, instrumentation, or extraction and derivatization technique. Provided column efficiency is carefully monitored and standard techniques are employed such difficulties can be avoided.

Some of the parameters causing actual profile change are as follows:

1. *Culture age*

It has been shown by Abel *et al.* (1963) that in batch grown *E. coli*, culture age has a considerable effect upon cellular fatty acid fingerprint. They examined samples representative of the "accelerated death" phase, "negative growth acceleration" phase and "logarithmic increase" phase. The latter samples appeared to contain rather more stearic than palmitic acid unlike the other samples. Also, rather less myristic, heptadecanoic and

TABLE XII
Analysis of amino-acids by GLC

Author	Derivative	GLC column	Temperature	Detector
Gehrke, Nakamoto & Zumwalt, 1969	Trimethylsilyl esters	3% OV-7 and 1·5% OV-22 on (100–120 mesh) Chromosorb G	75°C for 7 min, then to 225°C at 20°C/min	F.I.D.
Gehrke & Shahrokhi, 1966	N-trifluoroacetyl-n-butyl ester	0·75% DEGS and 0·25% EGSS-X on (60–80 mesh) acid washed Chromosorb W	67°C to 218°C at 3·3°C/min	F.I.D.
Moss, Lambert & Diaz, 1971	N-heptafluorobutyryl-n-propyl ester	3% OV-1 on (80–100 mesh) AW-DMCS-HP Chromosorb W	100°C for 5 min, then to 250°C at 40°C/min	F.I.D.
Moss, Thomas & Lambert, 1971	N-trifluoroacetyl-n-propyl ester	3% OV-1 on (60–80 mesh) Gas-chrom Q or 3% NGS HI-EFF-3CP on (60–80 mesh) Gas-chrom Q	—	F.I.D.

nonadecanoic acids were found. As Veerkamp (1971) has pointed out, changes due to ageing may be due, actually to change in pH. To avoid such effects, cultures of organisms in any one study should be grown to the same point on the growth curve; or to standard times of incubation, when different organisms growing at differing rates will possibly be in different phases of growth and therefore present different chemotaxonomic fingerprints.

2. Temperature

The effect of temperature on fatty acid profiles was shown by Marr and Ingraham (1962). In order to maintain the *physical* characteristics of lipids in the cytoplasmic membrane, micro-organisms grown at a lowered or raised temperature require to alter the *chemical* composition of the membrane's fatty acids.

Raised temperature would result in decreased viscosity and is compensated for, in part, by increasing saturation of fatty acids. For chemotaxonomic purposes, cultures must be grown at carefully controlled temperatures, preferably in a shaken water bath.

3. pH

This has been shown to affect the fatty acid fingerprints of *Streptococcus mutans* grown in a chemostat (Drucker *et al.*, 1975b).

Since many micro-organisms alter the pH of their environment, it is preferable to employ a chemostat or batch-fermenter where pH can be accurately controlled. This would clearly be impracticable where large numbers of organisms were being examined, and here a decision has to be made on whether to maintain the pH artificially, e.g. by growth in a well-shaken, chalk-containing or alkali titrated medium, or whether to allow the organism to establish its own environmental pH, which will partly be responsible for the micro-organisms characteristic fingerprint.

4. Carbohydrate source

Working with *Nocardia asteroides*, Farshtchi and McClung (1970) showed that if a glucose medium was substituted by one with glucose and amino-acids or glycerol or Dubos oleic albumin, then the fatty acid profile was affected qualitatively and quantitatively. Streptococci grown on 3% w/v sucrose in place of 1% w/v glucose also showed quantitative changes in fatty acid profile (Drucker *et al.*, 1974a).

Ideally, for chemotaxonomic purposes, a chemically defined medium ought to be employed. Failing that, different organisms should be grown in complex media either made up at the same time and stored in the freezer until required; or made up at different times with media of the same batch number.

5. *Vitamin and trace metal limitation*

Little appears to be known about the effects of such growth-limitations on chemotaxonomic fingerprints. Biotin is known to affect microbial biosynthesis of carboxylic acids (Wakil, 1960) and cellular composition (Gavin and Umbreit, 1965; Bunn, McNeill and Elkan, 1970). Magnesium-limitation has been shown to cause chemotaxonomic alteration in *Streptococcus* (Drucker *et al.*, 1974b).

To overcome these difficulties, nutritionally adequate media should be employed.

6. *Gaseous environment*

The influence of oxygen on fatty acid composition has long been appreciated. Recently the effect of aeration on chemostat-grown cultures of streptococci has been investigated (Drucker *et al.*, 1973) when it has been shown that the use of compressed air for sparging in place of 5% CO_2/95% N_2 results in increased proportions of unsaturated fatty acid being formed. Cultures therefore must be grown under rigorously defined gaseous conditions.

7. *Dilution rate*

This applies only to chemostat cultures and on the limited evidence available appears to have a quantitative effect upon fatty acid profiles. In practice there is a problem in deciding whether to control this effect by using a standard dilution rate, or whether to employ a rate which varies from one strain to another but represents a fixed fraction of μ max.

IV. PGLC OF WHOLE CELLS

A. Specialized instrumentation and analysis

Pyrolysis gas–liquid chromatography (PGLC) is a technique intended to overcome the sometimes limiting requirement of conventional GLC for volatility in the compounds to be separated, by thermally degrading prior to analysis. Whilst micro-organisms could be converted to volatile products by combustion, the products formed would be rather uniform, e.g. CO_2, H_2O, NH_3, etc. Pyrolysis uses heat in the absence of oxygen so that no net oxidation can take place and molecules are cleaved at their weak points, to produce smaller and fairly reduced, molecules. Originally PGLC was developed as a means of analysing paint plasticizers, rubber and other polymers. Its use in microbial chemotaxonomy was pioneered by Oyama and Carle (1967) and Reiner (1965).

Pyrolysis has sometimes been carried out as a separate step from GLC analysis, and sometimes has been carried out in a modified gas chromatograph so that pyrolysis products are not dropped but are swept immediately on to the column for analysis.

Three types of pyrolysis equipment are generally available:

(i) the heated furnace,
(ii) the heated filament,
(iii) the Curie-point pyrolyser.

The heated furnace has come in many forms generally being a tube packed with materials which sometimes have had catalytic capabilities. Samples which pyrolyse at very low temperatures have sometimes been analysed following pyrolysis at injection point heater temperatures.

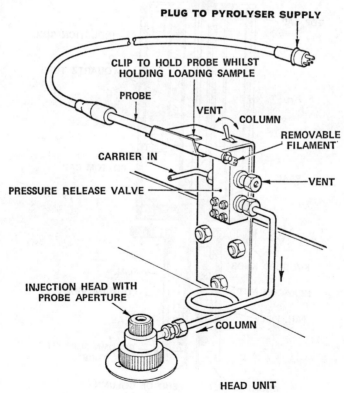

FIG. 20. Heated filament pyrolysis unit.

The most commonly used technique in microbiology has been the heated filament. This consists of either a platinum strip (Reiner *et al.*, 1971) or a nichrome filament on to which micro-organisms are placed. After the GLC has been allowed to equilibrate, the organisms are pyrolysed by means of a current supplied by the pyrolyser control unit, for a specified number of seconds. In the case of the heated filament, the pre-determined temperature is reached within 0·1 sec as energy stored in a capacitor is discharged through the filament (Fig. 20). The temperature varies with geometry of the filament, however, and this is theoretically disadvantageous.

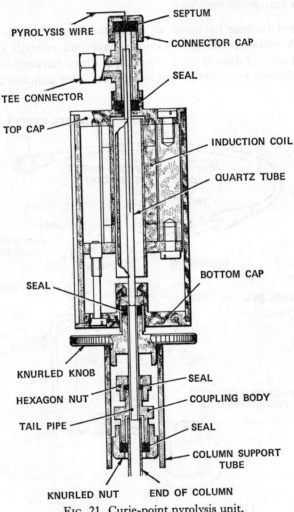

FIG. 21. Curie-point pyrolysis unit.

The Curie-point pyrolyser which has recently been used in microbial chemotaxonomy (Meuzelaar and In't Veld, 1972) is a much more sophisticated pyrolysis technique permitting far more accurate control of pyrolysis temperature. Time taken to reach the desired operating temperature is also very short. The Curie-point is reached when, at a particular temperature, a magnetic transition occurs from ferro-magnetism to paramagnetism and the metal wire can no longer be magnetized, or heat-up as a result of hysteresis losses, following the setting up of an alternating magnetic flux in the wire, by means of a high frequency (radio-frequency) current passed through a coil surrounding the wire, which in turn is surrounded by a quartz tube. The Curie-point will depend on the chemical composition of the wire which can, of course, be very accurately controlled. Nickel has a Curie-point of 358°C; iron, 770°C, a commonly used alloy of the two has a Curie-point of 510°C. Figure 21 shows diagrammatically a Curie point pyrolysis unit.

Regardless of the apparatus used for pyrolysis, the nature of the pyrolysis products changes markedly with changing pyrolysis temperature, and the technique is frequently classified, according to pyrolysis temperature as:

(i) mild pyrolysis,
(ii) medium pyrolysis,
(iii) high temperature pyrolysis.

In mild pyrolysis temperatures from 100–400°C are generally employed. This temperature will tend to break down polymers to monomers, and degrade the more thermolabile substances such as quaternary ammonium compounds.

Medium pyrolysis employs temperatures in the approximate range 500–800°C, which results in destruction even of fairly thermstable compounds; including primary products of pyrolysis which interact to form secondary products, which in turn react with one another and with primary products to form tertiary pyrolysis products. To minimize such reactions, pyrolysates are rapidly flushed on to the cooler analytical column by inert carrier gas. Medium pyrolysis yields data which vary from one laboratory to another even for the same micro-organism, although reproducibility within a laboratory may be satisfactory.

High temperature pyrolysis, in which temperatures up to 1200°C may be employed, results in drastic degradation of molecules which are converted into compounds of very low molecular weight.

The effect of temperature is well seen in the work of Lehmann and Brauer (1961) who showed that polystyrene was degraded to styrene and toluene at 425°C, and to those products plus ethylbenzene at 825°C. The choice of temperature for PGLC in microbial chemotaxonomy is therefore of the utmost importance.

TABLE XIII

Comparison of pyrolysis–GLC methods

Author	Sample analysed	Pyrolysis unit	Pyrolysis conditions	Column dimensions	Stationary phase	Oven temperature	Carrier gas	Detector
Reiner et al., 1971	Phenolized, lyophilized cells	Heated platinum ribbon	850°C/10 sec	Copper, 6 m; 0·17 cm i.d.	5% Carbowax 20M	30°–175°C at 12°C/min then isothermal	N_2; 40 ml/min	F.I.D.
Haddadin et al., 1971	Formalized suspension of cells	Heated filament	800°C/5 sec	Stainless steel, 3–7 m; 0·125 in. o.d.	20M Polyethylene glycol	50–145°C at 12°C/min; then isothermal	N_2; 20 ml/min	F.I.D.
Huis In't Veld et al., 1973	Cell envelope fraction applied to wire as a lyophilized suspension	Curie-point (Fe wire)	770°C/0·6 sec;1·5 kW; 1·1 μCi	Stainless steel, 32 m; 0·05 cm i.d.	10% Carbowax 20M	70°C isothermal/4 min; 70–140°C at 5°C/min; then isothermal	N_2; 4 ml/min	F.I.D.

At standard temperatures, differences in pyrogram reflect differences in chemical composition. Since alanylglycine yields ammonia and acetaldehyde at the same temperature as glycylalanine yields 2-methyl pyrrole, the complexity of products formed by pyrolysis of whole micro-organisms will be readily appreciated. In practice, a highly complex pyrogram is produced by micro-organisms and serves merely as a chemically ill-defined finger-print. The more precise analysis of the types described above for either fermentation product or cellular components is not possible.

In GLC chemotaxonomy a number of rival PGLC techniques have been devised. Cultural conditions have varied with the organism. Pathogens have been treated with 1% w/v phenol (Reiner et al., 1971) or 1·2% w/v formaldehyde (Haddadin et al., 1973). In Reiner's technique, treated cells are washed free of phenol, lyophilized and weighed out to the nearest μg then transferred to a platinum pyrolysis ribbon. In Haddadin's technique, washed cells were resuspended in a standard volume of distilled water and loaded on to the removable pyrolysis probe's filament by twice dipping into the suspension and drying in a stream of hot air. The individual operating conditions for PGLC for various workers are compared in Table XIII.

No work comparatively evaluating the various methods available has yet appeared in the microbiological literature and usually the choice of method probably will be determined by instrument availability rather than theoretical considerations.

One practical difficulty, frequently encountered, is poor reproducibility due to semi-pyrolysed material falling from the pyrolysis chamber on to the top of the GLC column. If this material is collected on a porous heat resistant filter and periodically removed, and if the top few inches of column packing are replaced, then columns can be used for several hundred analyses without deterioration in performance.

V. GLC-ANALYSIS OF BODY FLUIDS FOR THE PRESENCE OF MICRO-ORGANISMS

The traditional identification schemes employed by the medical micro-biologist generally are quite satisfactory. Problems may occasionally arise however if (a) the causative organism is non-viable by the time laboratory identification commences; or, (b) the time taken to correctly identify a slow-growing or atypical micro-organism is longer than the course of the patients' illness!

Considerable effort has been expended in attempts to avoid either or both the problems just mentioned by employing various rapid techniques such as fluorescent antibody, rapid fermentation techniques, differential light scattering patterns and, in the case of viruses, electron-microscopy.

Gas chromatographic techniques have been used, too, especially pyrolysis gas–liquid chromatography, which has been suggested as a technique
for rapid diagnosis of *Mycobacterium* by Reiner (1965). GLC of cellular
components also offers rapid diagnosis, although sample preparation times
are more time-consuming than for PGLC.

Another, novel, technique which is being explored is the GLC analysis of
body fluids, for traces of metabolites characteristic of particular infecting
organisms, using special high-sensitivity techniques. This approach is still
at a fairly experimental stage but it has been possible for some years, to
detect bacteraemia in urine (Mitruka *et al.*, 1970) caused by *Staphylococcus
aureus*, *Escherichia coli* K-12 or *Clostridium chauvoei*, gas chromato-
graphically. Similarly, Mitruka, Norcross and Alexander (1968) have used
GLC successfully in the detection of *in vivo* activity of infectious anaemia
virus.

Detection of micro-organisms in body fluids depends on production of
specific metabolites by the infecting organism. Various body fluids may be
analysed following suitable treatment as follows:

A. Blood samples

1. *Treatment of blood*

To 0·2 ml of heparinized blood, or serum or plasma is added 1·0 ml
1·09M tetra-methylammonium hydroxide and 0·5 ml ethanol. The reaction
is carried out in a glass conical centrifuge tube at 100°C for 1 h using a
marble as a stopper. After digestion, the sample is freeze-dried. The
products of the reaction are taken up in 0·5% ethanol and after suspension,
insoluble matter is sedimented by centrifugation. The supernatant is stored
at − 20°C until required for gas-chromatographic analysis (MacGee, 1968).
This method extracts such substances as long chain fatty acids.

An alternative treatment is that of Mitruka, Norcross and Alexander
(1968), in which 2 ml aliquots of serum are treated with 0·10 ml 5N HCl and
1·0 ml 0·2M HCl–KCl buffer of pH 2·0. Precipitated material is removed by
centrifugation and the supernatant fluid, containing any microbial meta-
bolite, is extracted three times with 10 ml diethyl ether. The ether extract is
analysed by GLC. Excess ether extract can be further analysed after
evaporation to dryness and then thrice adding methanol and drying down in
order to remove all traces of hydrogen chloride. After adding ammonium
hydroxide to pH 7·5, the sample is re-dried, then dissolved in 1·0 ml dry
pyridine by warming to 70°C/20 min, in a stoppered vial. The resulting
solution is silylated with 0·2 ml trimethylchlorosilane and 0·4 ml hexamethyldisilazane, and is centrifuged after 30 min/22°C, before being dried
with anhydrous sodium sulphate.

2. Gas chromatographic analysis of heated blood samples

The tetramethylammonium hydroxide digests of MacGee are analysed on a 6 ft stainless steel column of 4 mm i.d. packed with 0·5% benzoylated Carbowax 20M on acid washed glass beads (80–100 mesh).

The Carbowax 20M is benzoylated to improve its thermostability, by treatment with benzoyl chloride in pyridine.

Analyses are performed at 140°C isothermally for 5 min, then programmed to 230°C at 5°C/min and held at 230°C for 60 min. An injection temperature of 360°C is employed. An argon ionization detector was used originally (MacGee, 1968). Figure 22 shows chromatograms for extracts of micro-organisms.

The ether extracts of Mitruka are analysed by GLC on columns of 10% Carbowax 20M on Chromosorb (60–80 mesh) and on non-polar columns of 3% SE-30 on Chromosorb P (60–80 mesh). Temperature programming is employed to raise the starting temperature of 90°C to 225°C at 5°C/min. F.I.D. and E.C.D. detectors are used.

To analyse silylated compounds, columns of 15% EGS on Gas-chrom P (80–100 mesh) and 3% SE-52 on Gas-chrom P (80–100 mesh) are employed with temperature programming from 110–190°C at 10°C/min, using F.I.D. and E.C.D.

Typical chromatograms are shown in Fig. 23.

B. Urine samples

1. Treatment of urine

0·1 g tetramethylammonium hydroxide is dissolved in 1·0 ml urine which is then treated as described above in the MacGee method. A much more sophisticated treatment scheme is that of Mitruka which is shown in Fig. 24.

Mitruka's method is designed to permit detection of acid, neutral and basic compounds, including primary amines, urinary acids, urinary sugars and sugar alcohols, and nitroso-compounds.

2. Gas chromatographic analysis of treated urine samples

The MacGee samples can be analysed as described above in Section V.A, 2.

The samples heated according to Mitruka's scheme may be analysed as follows (Mitruka, 1974):

Urinary acid samples (Mitruka's Method) are analysed on a 12 ft column of 5% SE-30 on Gas-chrom P (80–100 mesh) with temperature programming from 90–260°C at 2°C/min, using FID.

Urinary sugars and *sugar alcohols* (Mitruka's Methods) are analysed on an identical SE-30 column but with temperature programming to 200°C only.

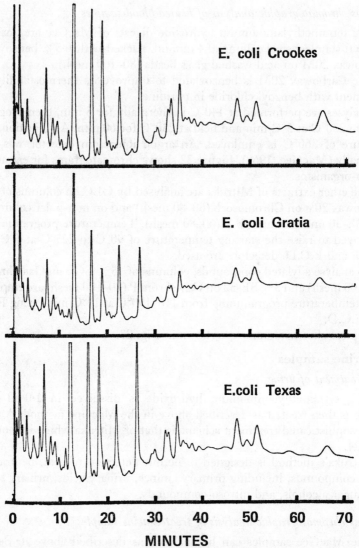

FIG. 22. Gas chromatogram of microbial culture extracts (after MacGee, 1968). Reproduced by kind permission of the Preston Technical Abstracts Company.

Amines and *nitroso-compounds* are analysed by the method of Brooks *et al.* (1972) (Mitruka's Method 6) on a 7·3 m column of 0·3 mm i.d. packed with 3% OV-1 on Chromosorb W (80–100 mesh) AW DMCS. After an initial isothermal period at 90°C/5 min, the temperature is programmed to 220°C at 5°C/min.

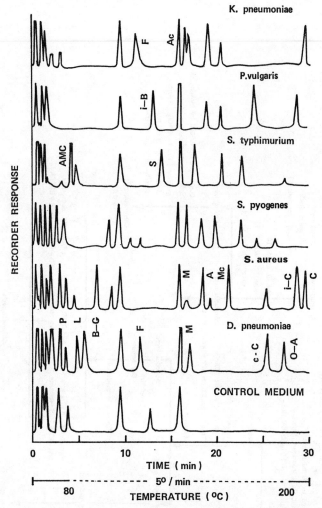

FIG. 23. Gas chromatogram of acids from cultures derived from patients' sera (after Mitruka, 1974). Key: (A) acrylic, (Ac) adipic, (AMC) acetoin, (B-G) butylene glycol, (*i*-B) *iso*-butyric, (C) citric, (*i*-C) *iso*-citric, (*c*-c) *cis*-aconitic, (F) fumaric, (L) lactic, (M) malic, (Mc) maleic, (O-A) oxalo-acetic, (P) pyruvic, (S) succinic. Reproduced by kind permission of John Wiley & Sons, London.

VI. GLC DATA ANALYSIS

The modern gas chromatograph can perform, in only minutes, analyses which formerly would have taken many hours, e.g. the analysis of fermentation products. Unfortunately, reduction in chemical analysis time high-

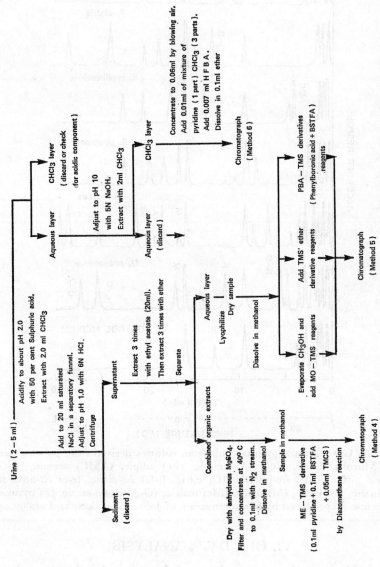

Fig. 24. Methods for gas chromatographic analysis of urine samples (after Mitruka, 1974). Reproduced by kind permission from *Applied Microbiology*

lights a new problem—the generation of vast amounts of data, which are time-consuming to analyse. The time taken, without electronic aids, to measure peak areas and relative retention times may be almost as great as the actual GLC analysis time. To overcome this problem, electronic integrators and computers have been successfully introduced into the modern gas chromatography laboratory.

An additional problem, in chemotaxonomy, is that comparison of fingerprints is largely subjective. Recently, attempts have been made to compare fingerprints on an objective basis using statistical techniques which not only are more satisfying academically than the traditional subjective approach, but make computer identification of unknown strains a real possibility.

Some of these recent developments are very briefly described below:

A. Integration of peaks

Accurate measurement of GLC peaks is essential for two reasons, (i) the retention time is characteristic of a compound, under standard conditions; the relative retention time of a peak compared with an internal standard is even more useful; (ii) the peak area is usually proportional to the amount of the compound present in an injected sample, within the limits of linearity of detector response, and linear range of the electronic amplifiers being used.

To measure retention times accurately, a stop watch can be used but this requires that personnel are able to constantly attend the gas chromatograph. Derivation of retention time from measurements of distance on a chromatogram are generally inaccurate. Electronic measurement of retention time is desirable.

To measure peak areas accurately without technical aids, an analysis requires constant supervision to ensure that peaks remain on scale; this is a wasteful use of personnel and is subject to a number of human and technical errors such as the following:

  (i) fatigue and wrong interpretation of the limits of peaks by personnel,
 (ii) inaccurate areas generated by triangulation and planimetry,
(iii) recorder faults such as lack of linearity, or excessive dead band.

1. *Electromechanical data handling*

A number of recorders are available with "disc integration". They produce two traces, one is a normal chromatogram, the other is a disc integrator trace, where a pen moves backwards and forwards over a distance of approximately 1 in., at a rate determined by peak area. The peak area must be calculated by counting the distance travelled by the integrator pen for each peak (Fig. 25). This form of integration is not ideal

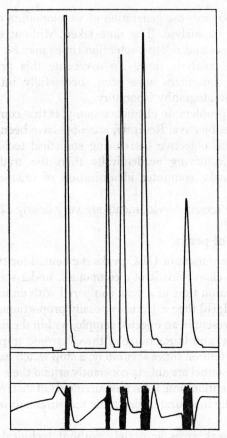

FIG. 25. Disc integrator trace, and relevant chromatogram.

for use in chemotaxonomy; it has a number of limitations. These limitations include lack of retention time measurement, lack of in-built correction for base-time drift, the need for subjective assessment of peak limits in the integrator trace, and lack of area print-out in digital form.

2. *Digital electronic integration*

This is more expensive than an electromechanical appliance but is more suitable for the frequently complex chromatograms obtained in chemotaxonomic studies. The amplified detector signal from a gas chromatograph is fed to the integrator which is usually connected in series with, and between, the amplifier and chart recorder. Amplifier output is fed to a voltage-to-frequency converter which produces pulses whose total number is proportional to peak area. Sophisticated electronics offer the following features:

  (i) A linear range of at least five decades. This in turn means that the more usual GLC amplifier which is linear over three decades will need to be replaced by a wide range amplifier (parametric amplifier), which can match the performance of the integrator.

 (ii) Variable peak bandwidth capability; for the accurate measurement of narrow peaks at the beginning of an analysis and broad peaks at the end of an analysis.

(iii) Variable slope detection capability; for ability to detect both steep and shallow peaks and differentiate them from electronic background noise.

(iv) Baseline correction facility; to allow for drifting baseline during an analysis.

 (v) Print out on paper or punched tape of peak retention time in seconds (usually four decades), peak area (usually seven decades) and other information (Fig. 13).

(vi) Additional features such as attenuation of output signals to chart recorder; inhibition of integration whilst solvent peaks elute; rejection of peaks below a minimum area; event markers which show start and finish of peak integration, etc.

Areas of fused or tailing peaks are generally formed by electronically "dropping a perpendicular line" to where the baseline is expected to be.

The most recent "generation" of integrators are the computing integrators which are in advance of the more elementary electronic integrator just described. They use more sophisticated logic and when calculating peak areas use a variety of very sophisticated corrections, e.g. for a skewed or tailing peak, an individual rider peak, a fused rider peak with tangential baseline correction, etc. A typical print-out illustrating such points is shown in Fig. 26.

In addition they can "double-up" as electronic calculators for determining percentage peak areas or concentrations, and may have an interface for use with a teletype or computer "on-line".

B. Computation of fermentation end-product data

The integrator print-out described above can be further analysed by computer. Integrator data can be fed into a computer (or a terminal) by typing in integrator print-out, manually or by first punching it on to card or tape or, preferably, by feeding integrator data from a punched tape on to magnetic disc via a fast punch-tape reader. The latest integrators, with suitable interfaces, can have a computer "on-line". By any of the means

118   D. B. DRUCKER

```
No. 54          51 ID
                 5 PW
                50 SS
                 5 BL
                30 TP
               500 MA
               100 T2
TIME          AREA
    4          1983 1
   21       6846441 BS
   76         10615 5
  131        166823 5
  170          9935 5
  235        255914 1
  278        461320 S
  347         11702 6
  375         27052 6
  433       1103237 6
  494        308377
  586          3268
  652         41811 2
  692         47486 3
  810        261907 1
  906        307808 S
 1142          6259 6
 1230         30589 7
 1691         19995
 2331         11130
 3195          8921
            9942573 T
```

FIG. 26. Computing integrator print-out. Key: (No. 54 51ID) injection number, (5 PW) peak width parameter, (50 SS) slope sensitivity parameter, (5 BL) baseline test parameter, (30 TP) tailing peak parameter, (500 MA) minimum area parameter, (100 T2) automatic doubling time for PW and SS, (TIME) retention time in seconds, (AREA) peak area, (1) peak receiving horizontal baseline correction, (B) input level > 1 V, (5) individual rider peak, (S) skewed or tailing peak, (6) fused rider peak given tangential baseline correction, (2) fused peak with trapezoidal baseline correction, (3) last peak in fused group to receive trapezoidal baseline correction, (7) last peak in fused group to receive tangential baseline correction. (T) total area.

just described, data can be filed on a magnetic disc until it is required for analysis. Actual computer analysis may consist of calculation of RRT (relative retention time to an internal standard), $\log_{10}$ RRT and ECN (equivalent carbon number—see below) with percentage peak area. Repeat analyses can then be further analysed to provide mean ECN, with S.D., and mean percentage peak area with ± S.D. Further computation can convert ECN values to tentative peak identifications and percentage peak areas to percentage composition, using correction factors, and either normalization or internal standard methods. For fermentation end-products, computation of the mean ECN and mean percentage area may be more than adequate. Computing integrators can calculate percentage composition as outlined above (Section VI.A).

With fermentation end-products, the use of ECN values can yield rather "untidy" data; since formic acid, like the lowest members of other homologous series is rather atypical, and does not have an ECN of unity when acetic = 2·0, propionic = 3·0.

An additional problem is that fermentation end-products represent a number of different homologous series, all of which should use different constants in the calculations of ECN values. Many workers will prefer to use tables of RRT values for peak identification.

C. Computation of cellular component data

Most of the published work is on cellular fatty acids. For chemotaxonomic purposes, computation of the mean percentage peak areas for substances of calculated mean ECN values is insufficient by itself and objective comparison of "fingerprints" by statistical analyses is desirable.

1. Computation of fatty acid composition

Integrator print outs of carboxylic methyl ester analyses are fed into a computer and percentage area, RRT (to methyl palmitate), $\log_{10}$ RRT and ECN are calculated. Furthermore, the computer can be programmed to scan repeat data sets to decide which peaks are the same, i.e. have same $ECN \pm 0·1$, then to calculate mean values, having first been provided with additional data such as the value of the ECN constant and the number of sets of data. For fatty acid ester ECN values where RRT is calculated relative to palmitate (16 carbon atoms):

$$ECN = 16·00 + \frac{(\log_{10} RRT)}{ECN \text{ coefficient}}$$

The ECN coefficient $\approx 0·129$ (Drucker, 1974). A typical print-out is shown in Fig. 27. During the initial computation, solvent peaks are excluded.

During averaging of peak areas, artefactual peaks are ignored, only peaks present in $> 60\%$ of analyses are included. Corrected mean percentage peak areas are calculated. Methods for further analysis of mean data are dealt with below (Section VI.E).

D. Interpretation of pyrograms

The advantages of PGLC for chemotaxonomic purposes have always been its speed and simplicity; the disadvantage has always been the difficulty of interpreting the pyrolysis-chromatogram. This is for two reasons; firstly, the use of an internal standard is not practicable and relative retention times cannot be obtained with reference to a known peak;

5248		483	
RRT	LOGRRT	ECN	AREA
.538302	−.268924	14.0006	4.43578
.730849	−.136146	14.9878	.223139
.809524	−9.17524E−2	15.3178	.270528
.875776	−5.75952E−2	15.5718	6.97379
1	1.34583E−6	16	33.9055
1.10766	.0444	16.3301	5.15133
1.50518	.177557	17.3201	.246363
1.63768	.214192	17.5925	1.27198
1.87785	.273611	13.0343	11.0358
2.11801	.325871	18.4228	24.7003
2.49068	.396249	18.9461	.49963
3.89855	.590797	20.3925	11.2859

FIG. 27a. Computer print-out of analysed methyl ester integrator print out. Top row lists injection number and palmitate retention time. Key: (RRT) retention time relative to palmitate, (LOGRRT) $\log_{10}$ RRT, (ECN) equivalent carbon number, (AREA) peak area expressed as a percentage of total area of peaks from ECN 13·5 up to 24·0.

SETS	RANGE	MEAN ECN	MEAN AREA	S.D.
3	.1	13.9917	4.64414	.216058
3	0	15.568	6.71918	.402772
3	0	16	33.8078	.348067
3	.05	16.3292	4.32401	.74817
2	.2	17.3223	.283391	5.13132E−2
3	.05	17.5945	1.21919	.103152
3	0	18.0254	10.9876	9.72139E−2
3	.15	18.4139	24.9911	.199919
3	.05	18.9443	.548145	8.13314E−2
3	0	20.3882	12.4755	1.00289

FIG. 27b. Computer print-out showing mean analyses for carboxylic methyl esters. Key: (SETS) no. of analyses including a compound, (RANGE) range of ECN values of averaged data.

secondly, the substances produced by pyrolysis bear only a very indirect relationship to the chemical composition of the micro-organisms being pyrolysed.

These difficulties have resulted in workers merely numbering the pyrolysis peaks and not seeking to identify them (Reiner *et al.*, 1972; Reiner and Ewing, 1968; Haddadin *et al.*, 1973).

Provided pyrolysis temperature, carrier flow rate, temperature etc. are all carefully monitored, it is possible to obtain reproducible retention times with which to "label" peaks for further, statistical, analysis.

E. Statistical analysis of GLC fingerprints

One of the criticisms of GLC-chemotaxonomy has always been that whilst it replaces subjective traditional tests with objective chemical analysis, the chemical data has then to be assessed. This assessment has frequently been highly subjective especially in cases where two strains of a micro-organism do not display qualitative chemical differences but differ only quantitatively. For a complete system of GLC-chemotaxonomic strain identification, the subjective data assessment step has to be eliminated. "Statistical analysis" of the data may be a misnomer since the various techniques employed have been used as loose "measures of association" rather than in a strict statistical sense. Some of the techniques which have been used are described below.

1. Coefficient of linear correlation

The author has found the Bravais–Pearson coefficient of linear correlation to be the most useful measure of association.

The coefficient

$$r = \frac{N\Sigma xy - (\Sigma x)(\Sigma y)}{\sqrt{[N\Sigma x^2 - (\Sigma x)^2][N\Sigma y^2 - (\Sigma y)^2]}},$$

x and y are variables; N = number of observations. x and y correspond to pairs of percentage peak areas for each peak in turn in a pair of chromatograms. A value of unity for r indicates complete correlation of two sets of data. A value of -1 indicates complete negative correlation; a value of 0 indicates no correlation. In order to decide which peaks in two analyses represent a "pair", it is necessary to use ECN values of standard retention times with pre-set ranges, for complete computerized analyses. This measure of association has been used with moderate success for comparing GLC chemotaxonomic data of some streptococci (Drucker et al., 1975a).

2. Spearman's coefficient of Rank correlation

This too, has been used as a loose measure of association and can be used with some degree of success for objectively analysing normalized GLC data (Drucker, 1974). One weakness of the method becomes apparent where finger-prints of two organisms both possess several peaks of only slightly different areas. False-ranking can result, yielding a slightly misleading value of $r_s = (1 - 6\Sigma d^2)/N(N^2 - 1)$, where d is the difference in rank for a pair of peaks and N is the number of peaks. Identical sets of GLC data will have r_s values of unity.

3. Extension of Fisher's method

This is an adaptation devised by Professor Violet Cane (personal communication).

$r = (\Sigma \sqrt{xi} \sqrt{yi})/100$, x and y are pairs of peak areas in two analyses; r is the cosine of the angle between the lines from the origin to the two points (representing two strains) on a hypersphere of radius 10. If the angle between two strains is $0°$, then the strains are identical and $r = 1$. If the angle between the strains is $90°$, the strains are as different as possible and $r = 0$. In practice, the author finds this is less satisfactory as a measure of association than the methods outlined above.

F. Computerized identification of micro-organisms by GLC-chemotaxonomy

Recent advances in instrumentation, such as the introduction of GLC-auto-injector systems and electronic integration of peak areas, have reduced, appreciably, the operator time required per sample analysis. Data sorting, which took a minor proportion of the sample analysis time, now accounts for the major part of operator time. If data analysis can be performed objectively, according to precise, mathematical rules, rather than subjectively, according to "experience", or "intuition", then it must be possible to use a computer for data analysis (Sections VI.A and VI.E). The author has, in fact, used a computer to carry out all the data analyses referred to in the preceding Section. These analyses effectively reduce chemical analyses for two strains at a time to a single measure of association which is analogous to the coefficient of similarity familiar to the numerical taxonomist (Sokal and Sneath, 1963).

By using the clustering and linkage methods of numerical taxonomy it is possible to sort organisms into clusters based on their association, instead of similarity coefficients.

There is no technical reason why micro-organisms or their extracts should not be injected on to a gas chromatograph and result in the production of clusters of similar phenotypes by computer, without further human intervention. In practice, it is convenient to split up computer programmes into suites of programmes which can be used one at a time, with the facility for operator intervention. In addition to this taxonomic application of computer techniques, there is no reason why a computer cannot be used to identify an organism by comparing its GLC-fingerprint to the fingerprints of "known" micro-organisms filed on magnetic disc. Figure 28 shows a print-out of comparisons between an unknown strain and a "library" of known strains. In addition to methods for performing the taxonomy and identification described above, Mitruka (1974) has suggested an interesting future use of computation in GLC-chemotaxonomy, samples will be analysed by GLC with a mass spectrometer "on the fly" which will examine the mass spectra of each compound as it emerges. The series of mass spectra produced would then be analysed by computer which would "perform an on-line computer matching of each mass-spectrum against a comprehensive reference file".

34.3853	1
47.978	2
16.8009	3
29.4634	4
27.2996	5
15.3831	6
6	15.3831

FIG. 28. Identification of an unknown micro-organism (*Streptococcus faecalis*) by comparison of its gas-chromatogram with chromatograms of known strains in a computer library. Key: (1, 2) *Streptococcus pyogenes*, (3, 4) *Streptococcus agalactiae*, (5) *Strep.* group C, (6) *Streptococcus faecalis*. First column lists "Hamming Distance", second column lists strain number. Last row lists most similar library strain to test organism and its Hamming distance.

Having obtained "measures of association" analogous to S-values in numerical taxonomy, further analysis is possible by employing the well-documented and well-tried cluster analysis techniques of the numerical taxonomist (Sokal and Sneath, 1963). Such techniques cluster together organisms of high measure of association so that two organisms which produce similar chromatograms will be grouped together, whilst organisms having dissimilar chromatograms will not be grouped together.

The marriage of computerized GLC-chemotaxonomy and numerical taxonomy has the advantage that raw data for cluster analysis is already provided by computer and therefore additional card-punching is not required.

At this juncture it seems likely that computation and numerical taxonomic techniques will increasingly be employed in the expanding field of GLC-chemotaxonomy.

REFERENCES

Abel, K., de Schmertzing, H., and Peterson, J. I. (1963). *J. Bact.*, **85**, 1039–1044.
Agate, A. D., and Vishniac, W. (1973). *Arch. Mikrobiol.*, **89**, 257–267.
Brian, B. L., and Gardner, E. W. (1967). *Appl. Microbiol.*, **15**, 1499–1500.
Brooks, J. B., Alley, C. C., and Jones, R. (1972). *Anal. Chem.*, **44**, 1881–1884.
Brooks, J. B., and Moore, W. E. C. (1969). *Can. J. Microbiol.*, **15**, 1433–1447.
Brooks, J. B., Kellogg, D. S., Thacker, L., and Turner, E. M. (1971). *Can. J. Microbiol.*, **17**, 531–543.
Bunn, C. R., McNeill, J. J., and Elkan, G. H. (1970). *J. Bact.*, **102**, 24–29.
Carlsson, J. (1973). *Appl. Microbiol.*, **25**, 287–289.
Chen, T. C., and Levin, R. E. (1974). *Appl. Microbiol.*, **28**, 681–687.
Cone, R. D., and Lechowich, R. V. (1970). *Appl. Microbiol.*, **19**, 138–145.
Cummins, C. S., and Harris, H. (1956). *J. gen. Microbiol.*, **14**, 583–600.
Doelle, H. W. (1967). *J. gas Chromatog.*, **5**, 582–584.
Doelle, H. W. (1969). *J. Chromatog.*, **42**, 541–543.
Drucker, D. B. (1970). *J. chromatog. Sci.*, **8**, 489–490.
Drucker, D. B. (1974). *Can. J. Microbiol.*, **20**, 1723–1728.
Drucker, D. B., Griffith, C. J., and Melville, T. H. (1973). *Microbios*, **7**, 17–23.

Drucker, D. B., Griffith, C. J., and Melville, T. H. (1974a). *Microbios*, **9**, 187–189.
Drucker, D. B., Griffith, C. J., and Melville, T. H. (1974b). *Microbios*, **10**, 183–185.
Drucker, D. B., Griffith, C. J., and Melville, T. H. (1975a). *Microbios Letters, in press*.
Drucker, D. B., Griffith, C. J., and Melville, T. H. (1975b). *Microbios, in press*.
Drucker, D. B., and Owen, I. (1973). *Can. J. Microbiol.*, **19**, 247–250.
Farshtchi, D., and McClung, N. M. (1970). *Can. J. Microbiol.*, **16**, 213–217.
Farshy, D. C., and Moss, C. W. (1970). *Appl. Microbiol.*, **20**, 78–84.
Gavin, J. J., and Umbreit, W. W. (1965). *J. Bact.*, **89**, 437–443.
Geddes, D. A. M., and Gilmour, M. N. (1970). *J. chromatog. Sci.*, **8**, 394–397.
Gehrke, C. W., Nakamoto, H., and Zumwalt, R. W. (1969). *J. Chromatog.*, **45**, 24–51.
Gehrke, C. W., and Shahrokhi, F. (1966). *Anal. Biochem.*, **15**, 97–108.
Haddadin, J. M., Stirland, R. M., Preston, N. W., and Collard, P. (1973). *Appl. Microbiol.*, **25**, 40–43.
Henis, Y., Gould, J. R., and Alexander, M. (1966). *Appl. Microbiol.*, **14**, 513–524.
Hill, L. R. (1966). *J. gen. Microbiol.*, **44**, 419–437.
Holdeman, L. V., and Moore, W. E. C. (1972). "Anaerobe Laboratory Manual", VPI & SU Anaerobe Laboratory, Blacksburg.
Hornstein, I., Alford, J. A., Elliott, L. E., and Crowe, P. F. (1960). *Anal. Chem.*, **32**, 540–542.
Huis, In't Veld, J. H. J., Meuzelaar, H. L. C., and Tom, A. (1973). *Appl. Microbiol.*, **26**, 92–97.
Ifkovits, R. W., and Ragheb, H. S. (1968). *Appl. Microbiol.*, **16**, 1406–1413.
James, A. T., and Martin, A. J. P. (1952). *Biochem. J.*, **50**, 679–690.
Kaneda, T. (1968). *J. Bact.*, **95**, 2210–2216.
Kersters, K., and de Ley, J. (1975). *J. gen. Microbiol.*, **87**, 333–342.
Kimble, C. E., McCollough, M. L., Paterno, V. A., and Anderson, A. W. (1969). *Appl. Microbiol.*, **18**, 883–888.
Lambert, M. A. S., and Moss, C. W. (1973). *Appl. Microbiol.*, **26**, 517–520.
Lee, S. M., and Drucker, D. B. (1975). *J. clin. Microbiol.*, **2**, 162–164.
Lehmann, F. A., and Brauer, G. M. (1961). *Anal. Chem.*, **33**, 673–676.
Littlewood, A. B. (1970). Gas Chromatography, 2nd ed. Academic Press, New York.
MacGee, J. (1968). *J. Gas Chromatog.*, **6**, 48–52.
Mangold, H. K., and Kammereck, R. (1961). *Chem. and Ind. Lond.*, pp. 1032–1034.
Marr, A. G., and Ingraham, J. L. (1962). *J. Bact.*, **84**, 1260–1267.
Martin, A. J. P., and Synge, R. L. M. (1941). *Biochem. J.*, **35**, 1358–1368.
Metcalfe, L. D., and Schmitz, A. A. (1961). *Anal. Chem.*, **33**, 363–364.
Meuzelaar, H. L. C., and In't Veld, R. A. (1972). *J. chromatog. Sci.*, **10**, 213–216.
Mitruka, B. M. (1974). "Gas Chromatographic Applications in Microbiology and Medicine." John Wiley and Sons, London.
Mitruka, B. M., and Alexander, M. (1969). *Appl. Microbiol.*, **17**, 551–555.
Mitruka, B. M., and Alexander, M. (1972). *Can. J. Microbiol.*, **18**, 1519–1523.
Mitruka, B. M., Jonas, A. M., and Alexander, M. (1970). *Infect. and Immun.*, **2**, 474–478.
Mitruka, B. M., Norcross, N. L., and Alexander, M. (1968). *Appl. Microbiol.*, **16**, 1093–1094.
Moss, C. W., Lambert, M. A., and Cherry, W. B. (1972). *Appl. Microbiol.*, **23**, 889–893.
Moss, C. W., Lambert, M. A., and Diaz, F. J. (1971b). *J. Chromatog.*, **60**, 134–136.

Moss, C. W., Lambert, M. A., and Merwin, W. H. (1974). *Appl. Microbiol.*, **28**, 80–85.

Moss, C. W., and Lewis, V. J. (1967). *Appl. Microbiol.*, **15**, 390–397.

Moss, C. W., Thomas, M. L., and Lambert, M. A. (1971a). *Brit. J. Vener. Dis.*, **47**, 165–168.

Orla-Jensen, S. (1919). "The Lactic Acid Bacteria." Copenhagen.

Orr, C. H., and Callen, J. (1958). *J. Am. Chem. Soc.*, **80**, 249.

Oyama, V. I., and Carle, G. C. (1967). *J. gas chromatog.*, **5**, 151–154.

Palo, V., and Ilková, H. (1970). *J. Chromatog.*, **53**, 363–367.

Prévot, A. R. (1948). "Manual de Classification et de Determination des Bacteries Anaérobies." Masson et Cie., Paris.

Reiner, E. (1965). *Nature, Lond.*, **206**, 1272–1273.

Reiner, E., and Ewing, W. H. (1968). *Nature, Lond.*, **217**, 191.

Reiner, E., Hicks, J. J., Ball, M. M., and Martin, W. J. (1972). *Anal. Chem.*, **44**, 1058–1061.

Reiner, E., Hicks, J. J., Beam, R. E., and David, H. L. (1971). *Am. Rev. resp. Dis.*, **104**, 656–660.

Rogosa, M., and Love, L. L. (1968). *Appl. Microbiol.*, **16**, 285–290.

Schlenk, H., and Gellerman, J. L. (1960). *Anal. Chem.*, **32**, 1412–1414.

Sokal, R. R., and Sneath, P. H. A. (1963). "Principles of Numerical Taxonomy." W. H. Freeman & Co., San Francisco and London.

Stahl, E. (1969). "Thin Layer Chromatography: A Laboratory Handbook." 2nd ed. Springer, Berlin.

Stoffel, W., Chu, F., and Ahrens, E. H. (1959). *Anal. Chem.*, **31**, 307–308.

Thoen, C. O., Karlson, A. G., and Ellefson, R. D. (1971). *Appl. Microbiol.*, **22**, 560–563.

Veerkamp, J. H. (1971). *J. Bact.*, **108**, 861–867.

Vincent, P. G., and Kulik, M. M. (1970). *Mycopath. Mycol. appl.*, **51**, 251–265.

Vorbeck, M. L., Mattick, L. R., Lee, F. A., and Pederson, C. S. (1961). *Anal. Chem.*, **33**, 1512–1514.

Wade, T. J., and Mandle, R. J. (1974). *Appl. Microbiol.*, **27**, 303–311.

Wakil, S. J. (1960). *Biochim. Biophys. Acta*, **41**, 122–129.

White, D. C., and Frerman, F. E. (1968). *J. Bact.*, **95**, 2198–2209.

Zlatkis, A., Oro, J. F., and Kimball, A. P. (1960). *Anal. Chem.*, **32**, 162–164.

CHAPTER IV

Transmission Electron Microscopy of Bacteria

KARL G. LICKFELD

Laboratory for Electron Microscopy at the Institute of Medical Microbiology, Clinical Hospital, University of Essen (GHS), D-4300 Essen, FRG

I. INTRODUCTION

Anyone who, like the author, has ever entered the wide field of bacteriological transmission electron microscopy (TEM) with pretended self-assurance derived from the fact that he has won his spurs somewhere in histological fine structure research, and who has nonchalantly applied his usual techniques to bacteria, only to fail pitifully a few days later—will know how far from the classical methods are some of the specific methods which must be used here. As soon as the unlucky person delves into the literature, it becomes clear that the mastery of bacteriological TEM starts with the fundamental studies on how to fix and embed bacteria, made by Antoinette Ryter and Eduard Kellenberger at the end of the fifties. Almost 20 years later, the number of specific methods has increased considerably and a situation has been reached where publications on methods for bacteriological TEM begin to confuse anyone starting research in this field. It is the aim of this contribution to present a basic selection of precise methodical instructions. Numerous variations of previously published methods remain unconsidered if they are in fact out of date or if the results achieved do not differ from those obtained with the original method or if they have merely been a result of good luck and have proved unrepeatable or unreliable. On the other hand, cryomicrotomy of bacteria is given careful consideration although this technique is still largely a blank area on the map of methods in microbiology. Its potentialities are, at present, scarcely conceivable.

II. SOME REMARKS ON THE TRANSMISSION ELECTRON MICROSCOPE

It is far beyond the scope of this contribution to give a comprehensive introduction to the principles of a transmission electron microscope. Several excellent books have been published recently, which are well suitable as a reference library, whenever detailed information on electron optics, on the principles and function of parts of the system, or hints on how to align an electron microscope are necessary (Agar *et al.*, 1974; Alderson, 1975; Hawkes, 1973; Meek, 1971; Reimer, 1967, Sjöstrand, 1967; Wischnitzer, 1973).

Besides text-books discussing fundamentals, the service manual of the instrument in use in the particular bacteriological laboratory has to be carefully borne in mind. Above all, the responsible scientist in the laboratory must of necessity become adept at operating the microscope, and should be able to do small repairs himself.

The more familiar technical assistants become with the instrument, the more effective will be the research work. Attendance at introductory courses such as those arranged for purchasers by all manufacturers of electron microscopes is most important and should never be omitted by default.

Each transmission electron microscope represents a complex unit. In principle, it consists of column, camera, vacuum system, high tension supply, low tension supplies, and control systems. The column is an evacuated tube, containing an electron source at the top, magnetic lenses, and a screen, at the bottom. Below the screen, the main camera is mounted. The specimen is inserted into the objective lens. Above the objective lens, two condenser lenses are located; together with the electron source, they represent the "lighting system", and the objective lens plus two or three projector lenses represent the "imaging system" (Fig. 1). Rotary symmetry governs the entire lens system.

As a consequence of the size of bacteria, relatively high magnifications of the microscope are the general rule; magnifications that are exceptions in

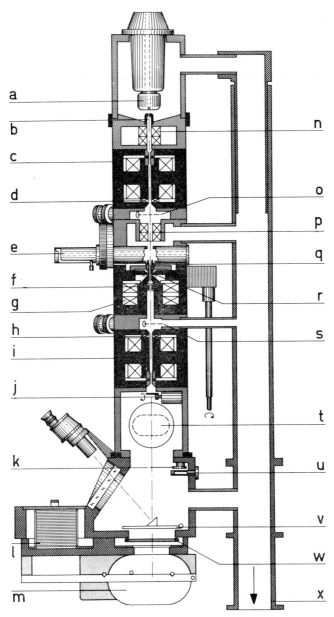

FIG. 1. Schematic longitudinal section of a typical Transmission Electron Microscope column ("EM 10"). Letters refer to some important parts as follows. a, cathode; b, anode; c, double condenser; d, condenser stigmator; e, specimen airlock; f, objective lens stigmator; g, objective; h, projector lens stigmator; i, double projector lens; j, shutter; k, central beam stop; l, plate camera; m, 70 mm roll film camera; n, beam alignment system; o, condenser aperture; p, beam tilting system; q, specimen cartridge; r, objective lens aperture; s, selector aperture; t, 35 mm camera or diffraction accessory; u, X-ray measuring tube; v, fluorescent screen; w, camera valve; x, pump column. (Reproduced by kind permission of Carl Zeiss, Oberkochen.)

other biological disciplines of practical electron microscopy. Survey pictures are rarely necessary. In routine work, most pictures are taken at image magnifications between about 20,000 : 1 and about 40,000 : 1. Fairly often fine structure details, e.g. membrane derivatives or specific granules, need to be examined carefully at relatively high magnifications of about 100,000 : 1 for several minutes, in order to understand a particular fine detail, or in order to find the so-called Scherzer focus or the physical focus, or to prepare for taking a critical focus series. From these requirements some fundamental properties of the instrument to be used for bacteriological electron microscopy can be deduced:

(a) The instrument needs to be equipped with an effective anti-contamination device, so that the duration of examination of a certain specimen detail can be prolonged as required.

(b) The instrument needs to be equipped with a double condenser to achieve a bright enough image on the screen and, at the same time, a sufficiently low specimen load by the electron beam.

(c) The instrument needs to be carefully adjusted, particularly with regard to compensation for axial astigmatism. Fine structural details of bacteria, as may be achieved by some of the methods described elsewhere in this contribution, are so small that even traces of axial astigmatism severely impair the usefulness of a micrograph.

Astigmatism preferably is corrected with the help of the so-called minimum contrast method at high magnification, which, for the beginner, may be the second step of adjustment after a first one, which is achieved by the use of a holey film.

In bacteriological TEM, the permanent use of thin foil apertures in the objective lens is highly recommended to ensure the highest possible stability of the corrected astigmatism.

III. SPECIMEN PREPARATION

At the end of the thirties, bacteria were one of the first specimens examined, when TEM began to be applied as a research tool in sciences involving structural examination. Of course the size and shape of bacteria could easily be determined but it proved almost impossible to see any details at all in the interior of the cells, not to mention resolving fine structure.

The collapsed matter of the intact cells, although chemically fixed, masked the macromolecular architecture, and the thickness of specimens was too great to allow of relatively high resolution. Methods of specimen preparation had to be developed to enable the bacterium to be sectioned into thin enough slices. Twenty years passed before thin sectioning could be performed successfully. From sectioned† bacteria, micromorphology had to be reconstructed. Only recently, with the help of high voltage transmission electron microscopy (HVTEM) and by means of the so-called "critical point method", has it become possible to demonstrate successfully undisturbed fine structure in intact cells (Porter *et al.*, 1974). This technique may well be helpful for resolving some of the remaining topological questions in bacteriology.

A. Negative staining

The main applications of negative staining are to be found in the demonstration of viruses and biological macromolecules. In addition, negative staining plays a significant role in high resolution bright- and dark-field transmission electron microscopy. But there are, occasionally, situations in bacteriological TEM, when negative staining is very helpful (e.g. demonstration of bacteriophages (Fig. 2)), apart from the necessity of

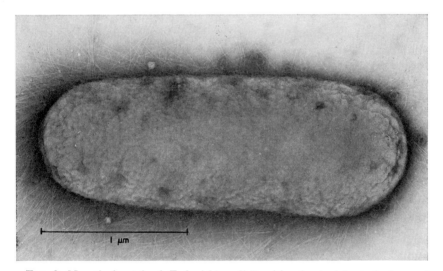

FIG. 2. Negatively stained *Escherichia coli* B with adsorbed bacteriophages λ. The cell surface appears wavy. Numerous flagella are clearly demonstrated.

† Some authorities favour the use of the terms "slice" and "slicing" rather than "section" and "sectioning" (see Elias, Hennig and Schwartz. Stereology: Applications to Biomedical Research. *Physiological Reviews*, **41**, 158–200 (1971)). The more usual terms "section" and "sectioning" are used in the present Chapter at the insistence of the Editor and contrary to the wishes of the author on the grounds that they are established English usage unlikely to change, that their meaning is quite clear and that the usage is in line with that applied in the rest of the Series.

using this technique, when components of disintegrated bacteria have to be studied (e.g. ribosomes).

For negative staining, a 2% solution of phosphotungstic acid (w/v) in distilled water is used. It is adjusted to pH 7·2 with 1N KOH.

Preparation is performed on carbon film-coated specimen grids. The thickness of the carbon film must not exceed about 20 nm, since otherwise contrast is impaired. If carbon coating is performed on collodion or Formvar film, the films should be made so thin that they disintegrate under electron bombardment. Otherwise specimen grids should be coated directly with pure carbon.

If carbon coating is omitted, it must be recognized that the heating of negatively stained specimens by the electron beam causes sagging of the respective square of the specimen grid. This, obviously, is the more marked, the higher the beam current. The fundamental rule, to first stabilize a loaded specimen grid by irradiation from the overfocused condenser system for some minutes before it is focused, is of utmost importance here. This rule is of course also valid for carbon film-coated specimen grids.

If focusing is performed on sagging films, unsharp pictures will be obtained, because sagging is reversible and therefore almost or completely cancelled as soon as screen brightness is reduced to take a picture. Thus focusing essentially has to be performed at the lowest possible beam current giving adequate screen brightness. An image intensifier is extremely helpful for focusing of negatively stained preparations.

The negative staining procedure is carried out as follows. A drop of the specimen suspension is placed on the specimen grid with a Pasteur pipette. The pipette is held vertically. The drop is left standing on the grid for 2 min. It is removed from the side with a piece of filter paper. A thin film of

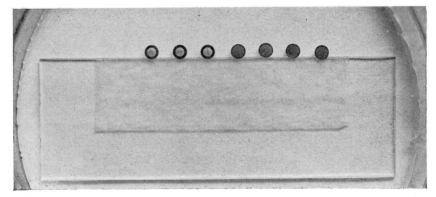

FIG. 3. The simple device that facilitates negative staining procedures, comprising a microscopic slide and a piece of tape, adhesive on both sides.

fluid should remain on the specimen grid. Immediately afterwards, a drop of the staining solution is added by means of (another) Pasteur pipette. The stain should act for only 20 to 40 sec and is then removed with filter paper. The more completely the stain is removed, simply by touching the drop somewhere at its rim, the more effective will be the enveloping of the object by stain.

A very simple device will prove extremely helpful for negative stain preparations (Fig. 3). A piece of double adhesive tape is attached to a microscope slide. Specimen grids are lightly attached to the edge of the tape with their edge only. Manipulations can then safely be performed, undisturbed by annoying capillary phenomena.

The loaded specimen grid is immediately put into the microscope. It should be left there for some minutes, before the cathode is activated. The specimens have to be vacuum dried before being irradiated. Otherwise wet stain will boil under electron bombardment, in this way impairing biological fine structures.

If a genuine liquid culture specimen is used for negative staining, difficulties may arise from the broth components. If this is so, the loaded specimen grid is rapidly rinsed once with distilled water after elapse of the adsorption time and before staining and insertion into the microscope.

Adsorbed bacteria may be chemically fixed before being negatively stained. This is done by holding the loaded specimen grid into the neck of an osmium tetroxide fixative bottle for a few seconds.

B. Microtomy techniques

1. Conventional microtomy

(a) *Cultivation of bacteria for TEM.* In general, pure cultures are stored on or in appropriately selected solid nutrient media.

A platinum loop-full of bacteria is used to innoculate 30 ml of freshly prepared liquid medium at room temperature and the culture incubated overnight in a gyrotory water bath shaker at the appropriate temperature.

From the overnight culture, sufficient drops are added to 30 ml of freshly prepared liquid medium at room temperature to give an optical extinction of about 0·03. This liquid culture is incubated in a gyrotory water bath shaker until the intended growth level is reached.

Any nutrient medium may be used, but semi-solid media should be avoided, since difficulties then inevitably arise in connexion with embedding procedures.

In many cases, Difco Bacto-Tryptone, B 123, is an excellent nutrient for chemo-organotrophic bacteria. It is equally well suitable for Gram-negative and Gram-positive bacteria. Merck Standard I medium has proved to be especially suitable for Gram-positive bacteria.

In any case, both the Difco manual (Anonymous, 1953) and the data given in the ATCC manual (Anonymous, 1972) are very helpful for the selection of the best suited nutrient medium. The DSM manual (Anonymous, 1974) and the Merck manual (Anonymous, 1968) should also be consulted.

If synthetic media have to be used for chemo-organotrophic bacteria, those should be preferred, which do not contain PO_4^{2-} ions. The presence of PO_4^{2-} ions during prefixation leads to the formation of precipitates in the cells.

For lithotrophic bacteria, specific nutrient media are essential. One has to rely on primary literature here or optimum growth conditions have to be determined experimentally. The details presented in Volumes 3A and 3B of this Series contain useful information regarding media selection.

The growth curve of a bacterium should be measured before electron microscopic studies begin. The fine structure of bacteria depends on many factors, among which the position of a culture on its growth curve is particularly important. Details of methods for measuring growth curves of bacteria are given in Volumes 1 and 2 of this Series (see also Oberzill, 1967).

(b) *Chemical fixation.* Chemical fixation of bacteria is always a three-step procedure, of which steps one and two are variable as to the fixatives used; step three is usually performed with uranyl acetate. It has become usual practice to use the terms "pre-", "main-" and "post-fixation" for these steps.

(i) *Pre-fixation and main-fixation with osmium tetroxide.* This is the most commonly used fixation technique and is generally known as "osmium fixation under RK conditions" (Kellenberger *et al.*, 1958; Kellenberger and Ryter, 1964; Ryter *et al.*, 1958).

A 1% solution of OsO_4 (w/v) in a modified Michaelis buffer, "RK buffer", has to be provided. The modification concerns the addition of Ca^{2+} ions, which prove to be indispensable.

The following stock solutions are prepared:

Solution A: 9·714 g sodium acetate ($CH_3COONa.3H_2O$)
14·714 sodium barbiturate ($C_8H_{11}N_2NaO_3$) in 500 ml boiled distilled water
Solution B: 0·1N HCl
Solution C: 8·5% sodium chloride (NaCl) in boiled distilled water (w/v).
Solution D: 1M calcium chloride ($CaCl_2$) in boiled distilled water. ($CaCl_2$: 11 g/100 ml; $CaCl_2.2H_2O$: 14·7 g/100 ml; $CaCl_2.6H_2O$: 22·8 g/100 ml)

All solutions should be kept in ground glass stoppered bottles. Solution A tends to become contaminated with fungi (e.g. *Fusarium* sp.). Storage at low temperature is not necessary.

Good preservation of bacterial fine structures is obtained when the pH of the fixative is 6, but different pH values may on occasion be necessary. The following table will allow for all eventualities.

TABLE I

Composition of RK buffer for different pH

pH	Solution B (ml)	Boiled distilled water (ml)
3·9	13	5
4·1	12	6
4·3	11	7
4·7	10	8
4·9	9	9
5·3	8	10
6·1	7	11
6·8	6·5	11·5
7·0	6	12
7·3	5·5	12·5
7·4	5	13
7·7	4	14
7·9	3	15
8·2	2	16
8·6	1	17
8·7	0·75	17·25
8·9	0·5	17·5
9·2	0·25	17·75
9·6	—	18

To each of the above given mixtures are added
of solution A: 5 ml,
of solution C: 2 ml,
of solution D: 0·25 ml.

It is advisable to check the pH of the freshly prepared RK buffer to ensure that mistakes have not been made and to control the quality of the ingredients used.

The ionization constant of OsO_4 is extremely low (Millonig and Marinozzi, 1968). Thus the osmotic concentration of a buffer as a vehicle for OsO_4 remains to all intents and purposes unchanged.

The vial, which contains OsO_4, should be carefully cleaned with cold tap and distilled water, and dried with filter paper before being opened. It should be broken in the bottle, which contains the appropriate amount of RK buffer. A cut around the vial, made with a small vial saw, considerably facilitates breaking. Some experience is necessary to make the cut deep enough so that a small lateral force only is needed to break the vial.

Safety goggles must be worn during these manipulations, since OsO_4 vapour may cause dangerous loosening of the cornea.

It takes a few hours to dissolve the OsO_4 crystals. The ground-glass stoppered bottle must be stored in the dark and should be wrapped with aluminium foil.

For pre-fixation, OsO_4 fixative is added to the liquid culture in such a proportion that a 0.1% final concentration (v/v) results; for instance 0.3 ml of fixative to 30 ml broth. The culture flask is shaken vigorously and left standing for 15 min at room temperature. The broth immediately starts to become blackish brown.

After 15 min, the contents of the culture flask are transferred to a centrifuge tube and centrifuged at 4000 g for 5 min. The supernatant is discarded. The pellet is immediately covered with 1 ml OsO_4 fixative and 0.1 ml of freshly prepared, amino-acids-containing nutrient medium, for instance Difco Bacto-Tryptone, B 123. Instead of this, an amino-acids mixture can be used, for instance Difco Bacto-Casamino Acids, B 231. If a synthetic nutrient medium was used for cultivation, amino-acids are absolutely essential.

By use of a thin, pointed glass rod, the pellet is carefully detached from the tube wall and then vigorously vortexed. Afterwards the tube is closed with Parafilm or with aluminium foil and left standing at room temperature for 10 ± 4 h. This is the main-fixation period.

Obviously, it is often advisable to start main-fixation in the late afternoon, in order to use time economically.

(ii) *Pre-fixation with glutaraldehyde and main-fixation with osmium tetroxide*. It has been shown recently that even very fragile protein structures can satisfactorily be fixed by the sole use of osmium tetroxide (Lickfeld, 1976b). On the other hand, it is well known that, for instance, protoplasts cannot be preserved intact by means of osmium fixation. The general rule that optimum preservation of living matter is probably obtained by use of two additive chemical fixatives, of which one preferably cross-links with protein and the other one with lipids, gains validity here. Successive fixation with glutaraldehyde and osmium tetroxide is an obvious procedure.

Contrary to OsO_4, glutaraldehyde exerts a relatively high osmotic pressure. It therefore changes the osmolarity of its vehicle. Furthermore, it tends to decompose. Only freshly prepared glutaraldehyde should be used

(Hayat, 1970) or, even better, glutaraldehyde which is supplied under nitrogen in sealed vials.

Glutaraldehyde concentrations between 2·5% and 5% (v/v) have successfully been used for pre-fixation of bacteria. Both RK buffer and cacodylate buffer are used as vehicles. RK buffer is prepared as described in (i) of this Chapter, for cacodylate buffer, the following stock solutions are prepared.

Solution A: 42·8 g sodium cacodylate [Na{(CH₃)₂AsO₂}.3H₂O] in 1000 ml boiled distilled water
Solution B: 0·05N HCl

According to the following table, solutions A and B are mixed to obtain a cacodylate buffer with the desired pH.

TABLE II

Composition of cacodylate buffer for different pH

pH	Solution B (ml)
5·0	23·5
5·2	22·5
5·4	21·5
5·6	19·6
5·8	17·4
6·0	14·8
6·2	11·9
6·4	9·2
6·6	6·7
6·8	4·7
7·0	3·2
7·2	2·1
7·4	1·4

To each of the above volumes of solution B are added 25 ml of solution A and then boiled distilled water to a final volume of 100 ml.

If, for instance, a 2·5% glutaraldehyde cacodylate buffer at pH 6 is required and a 10 ml vial of 8·5% glutaraldehyde in water is available, the fixative is prepared as follows.

Twenty-five millilitres of solution A are added to 14·8 ml of solution B and the mixture diluted to 100 ml with water to arrive at a pH 6 cacodylate buffer solution. The glutaraldehyde at hand must be diluted actual concentration/desired concentration times. This is 8·5/2·5 = 3·4. The contents of the vial are filled up with solvent to a final volume of 10 × 3·4 = 34 ml. Taking this into account, the actually required proportions of the buffer components are calculated as follows.

One hundred millilitres of a pH 6 cacodylate buffer solution contain 25 ml of solution A, 14·8 ml of solution B, and $100 - (25 + 14·8) = 60·2$ ml of boiled distilled water (solvent C). The A : B : C proportion of $25 : 14·8 : 60·2 = 1·69 : 1 : 4·07$ must also be present in the 34 ml portion of the fixative.

Here, the total volume of water is $C' = (34 \times C)/100 = (34 \times 60·2)/100 \approx 20·47$ ml. Accordingly, the volume of solution B is $B' = (B \times C')/C = (14·8 \times 20·47)/60·2 \approx 5·03$ ml. The volume of solution A is $A' = (A \times C')/C = (25 \times 20·47)/60·2 \approx 8·5$ ml. Of the 20·47 ml of water, 10 ml are added by the vial contents. Therefore, the desired fixative is composed of 8·5 ml of solution A, 5·03 ml of solution B, 10·47 ml of boiled distilled water, plus the vial contents. Thirty-four millilitres of 2·5% glutaraldehyde (v/v) fixative in a pH 6 cacodylate buffer solution are at hand.

For glutaraldehyde stock solutions of other concentrations (e.g. 25 or 70%) and/or other buffer solutions, similar conversions must be made.

For pre-fixation, equal amounts of liquid culture and fixative are combined. The flask is vigorously shaken and then left standing for 20 min at room temperature.

It is still uncertain whether pre-fixation should be performed at 0°C or 4°C (Hayat, 1970) or, as proposed here, at room temperature. Our own experiments point to shrinkage of cells when pre-fixation takes place at the temperature of melting ice and even at 4°C. The crucial point seems to be thorough mixing of culture fluid and glutaraldehyde fixative right at the start of pre-fixation.

Since glutaraldehyde and osmium tetroxide react, thorough washing has to be carried out between pre- and main-fixation. The pre-fixed material is transferred to a centrifuge tube and centrifuged at 4000 *g* for 5 min. About 5 ml RK buffer at pH 6 are added to the pellet, and pellet and buffer are vigorously vortexed. The suspension is centrifuged and again 5 ml RK buffer are added. By vortexing a homogeneous suspension is produced, which is shaken continuously for 1 h. Another centrifugation completes the washing procedure. If washing is inadequately carried out, precipitates will be found in the bacteria.

Main-fixation is performed as described in (i) by use of 1% OsO_4 (v/v) in RK buffer with the addition of amino-acids-containing broth or pure amino-acids.

Double fixation with glutaraldehyde and osmium tetroxide proves to be especially suitable for Gram-positive bacteria, whereas Gram-negative bacteria often show a less good preservation of fine structures. This different reaction is caused by different cell wall and cytoplasmic membrane compositions. It has been shown that the semipermeability of the cytoplasmic membrane is immediately abolished under the impact of pre-fixation osmium tetroxide. The relatively high osmotic pressure of the bacterial cell (order of magnitude: 10 atmospheres) is thus abruptly released. This effect may exert serious secondary actions if, for instance, intracellular phages and their fragile precursors are to be studied.

Phase-contrast light microscopy is a powerful and indispensable tool for the control of any fixation process. The lighting system of the instrument has to be carefully adjusted in combination with high aperture objectives (aperture $\geq 1\cdot3$), in order to ensure work at the limit of resolution (see Quesnel, this Series, Vol. 5A). Cytoplasm and nucleoplasm of bacteria can clearly be differentiated as well as certain granular inclusions.

(iii) *Fixation measures to suppress artificial membrane systems.* Evidence has been presented that at least the majority of mesosomes so far described are artefacts (Fooke-Achterrath *et al.*, 1974). Artifacts of this kind, "technikosomes", can reliably be avoided when pre-fixation is carried out on chilled samples at a temperature of 1°C to 4°C, using a pre-cooled fixative.

Liquid cultures of bacteria to be chemically fixed according to this technique, are cooled down in a cooling water bath. Precise thermometers are a necessity for temperature control of bath, samples and, possibly, fixative. Duration of pre-fixation remains as it was then carried out at ambient temperature.

After pre-fixation, samples are pelleted in a cooled centrifuge. Main-fixation is carried out as described in (i) at room temperature.

The reservations concerning pre-fixation in the cold, as given in (ii), are, of course, valid here too. But there is no method at hand yet to avoid this dilemma other than continuously shaking samples during pre-fixation.

(iv) *Pre-fixation with formaldehyde and main-fixation with osmium tetroxide.* If the aim of TEM studies is the demonstration of intracellular phages (e.g. T 4) and their quantitative measurement in thin sections (Séchaud *et al.*, 1959), neglecting good preservation of the bacterial cell, pre-fixation with formaldehyde is recommended. An advantage of this method is that no partial lysis is induced.

The concentration of commercially available formaldehyde varies between about 20% and 40%. It has to be of "EM grade", that is extremely pure and free from methanol.

The final concentration of formaldehyde in the liquid culture must be 3% (v/v). If, for instance, a 37% stock solution of formaldehyde is available and 30 ml of culture are to be pre-fixed, 2·64 ml fixative are added. The flask is vigorously shaken or vortexed and then left standing for 20 min at room temperature.

After centrifugation, main-fixation is performed under RK conditions, as given in (i).

(v) *Pre-fixation with higher concentrations of osmium tetroxide and main-fixation with osmium tetroxide.* The concentration of a fixative is one of the factors influencing its penetration rate. Therefore it was an obvious step to try to improve pre-fixation simply by use of final concentrations of osmium tetroxide higher than those originally recommended.

The maximum solubility of OsO_4 in cold water is 5·7% (w/v). So, in practice, pre-fixation final concentrations of this fixative cannot be made higher than a few per cent. In addition, it has to be taken into account that OsO_4 is a very expensive chemical.

This essentially leads to drastic reduction of liquid culture volumes when pre-fixation is to be performed with relatively high concentrations of OsO_4, (ca. 1% to 3% (w/v)). This, in turn, requires special precautions and techniques to obtain large enough pellets of material (see p. 143).

The use of pre-fixation final concentrations higher than 0·1% (v/v) so far has been restricted to studies on bacteriophage morphology and morphogenesis.

It has been shown (Lickfeld, 1976b) that an increase of final concentration by a factor of 2, from 0·1% to 0·2%, dramatically improved the appearance of "pregnant" *Escherichia coli* B cells containing bacteriophage λ. 1% stock solutions of OsO_4 in RK buffer are used for pre-fixation final concentrations of 0·2 to about 0·5%.

If experiments with higher concentrations promise to lead to better results, highly concentrated OsO_4 stock solutions must be prepared. For instance, 4 ml of a 2% fixative are added to 4 ml of liquid culture to give a final concentration of 1%. This has successfully been applied to bacteriophage T 4 (Okinaka, 1971) and will be equally good for other T-even bacteriophages.

Here again, main-fixation is performed as given in (i).

(c) *Suspension in inert carrier.* The colour of samples changes from blackish brown to black during the course of main-fixation. It is worth mentioning here that this blackening is, in fact, mainly caused by reactions between OsO_4 and bacteria. But, in addition, reactions take place between fixative and nutrient medium as can easily be demonstrated. This side effect yields relatively small amounts of electron opaque granules with diameters of about 10 nm, which tend to adhere to cell walls of intact cells, to cytoplasmic membranes of fully or partly disintegrated cells or, in general, to fragments of biological material. The interpretation of granular fine structures resting upon specimens needs to be done with care and after due consideration of the possibility of such artifacts.

After completion of main-fixation, samples are centrifuged again at 4000 *g* for 5 min. The black supernatant is discarded and the dark brown to black pellet is resuspended in about 5 to 10 ml of RK buffer using a pointed glass rod and/or shaking or vortexing. A second 5 min centrifugation follows. The pellet, that arises from this step needs special treatment for further handling. It has to be embedded in agar. This is an ingenious trick. If it is not embedded, the specimen will fall to pieces during the following manipulations.

The agar used must be free from poisonous additives. Agar intended for microbiological purposes is normally suitable.

Agar (0·4 g) is added to 20 ml RK buffer a few hours before it is needed. This soaking facilitates later solution of agar. For this, a suitable Erlenmeyer flask is recommended. It should be closed with aluminium foil. If main-fixation is carried out overnight, soaking should start the same evening.

About half an hour before embedding is to be performed, the agar-containing flask is put into an appropriate beaker with distilled water. This is brought to the boil and the agar-containing flask remains in the boiling water until the agar is completely solved. Solution is complete when the agar looks homogeneously clear. It is then transferred to a temperature stable 47°C water bath. It must not be used before the temperature of the agar has fallen to 47°C.

The 47°C water bath should be large enough to take a submerged test-tube rack. This should be provided with holes of various diameters, so that test-tubes and centrifuge tubes can be held.

A portion of the 47°C agar is added to an immersed test-tube and a 1 ml pipette placed in this portion.

After discarding the supernatant, the pellet-containing centrifuge tube is placed, upside-down, on filter paper. Then the interior of the tube is carefully dried, starting immediately above the pellet, using rolled filter paper, that just fits into the tube. The cleaned tube is put into the submerged tube

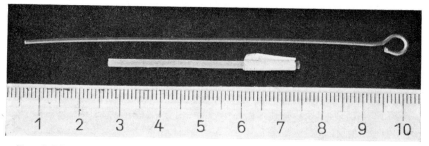

FIG. 4. The tool necessary for the "micromethod", comprising a piece of stainless steel wire with a diameter of about 1 mm and a piece of plastic tubing just fitting over the wire. One end of the tubing is marked with coloured tape to avoid confusion when several "micromethod" tools are used for different samples. Scale in cm.

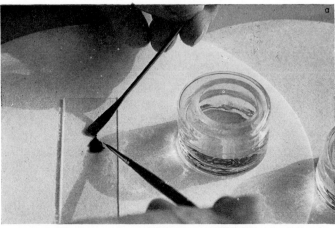

FIG. 5. Preparation of chemically fixed bacteria after suspension in agar. (a) By means of a small spatula, the specimen-containing strip has just been transferred to the microscope slide. The glass container holds a solution of uranyl acetate in RK buffer. The small brush has been used to put some drops of it on the slide. (b) With the help of a stiff razor-blade, the central part of the specimen-containing strip is exposed.

rack and left standing there for about 2 min until it is warmed up to 47°C.

Warm agar is then added to the equally warm pellet. The proportion pellet volume/agar volume proves to be fairly critical and should be about 1 : 1. If too little agar is added, the specimen becomes brittle. If too much agar is used, it will be difficult to find "cell rich" zones in the block later.

A pointed glass rod with a shaft diameter of about 3 mm is used for stirring the mixture of fixed bacteria and liquefied agar. At the start it looks piebald, later more or less homogeneous. The turbulent water in the water bath may lift the tube. It must therefore be held with one hand while the

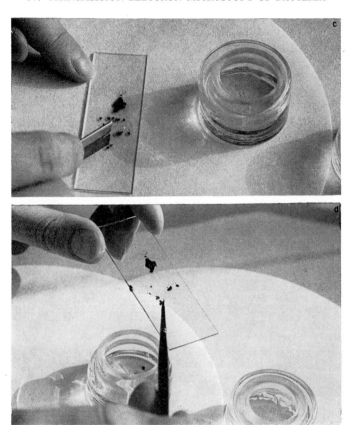

FIG. 5. (c) The central part has been cut into very small cubes. (d) The cubes are transferred into uranyl acetate solution with the help of a small brush.

other stirs. When a homogeneous state is reached, the tube is removed from the water bath, dried externally with filter paper and placed, upside-down, on filter paper. It is allowed to cool to room temperature. It may safely be left standing for 5 min or longer, because the wet chamber effect prevents the specimen from drying out. So a series of tubes can be prepared before the next preparational step starts.

In many cases, 30 to 50 ml cultures are the rule and they are handled as described above. If, however, volumes of less than 2 ml have to be pre-pared, another technique is recommended (Kellenberger et al., 1972). Small 2 ml polyethylene vials with a hinged cap are used. These fit into the rotors of small table-top centrifuges. After main-fixation and centrifugation, a special tool is used to disperse the pellet in the conical tip with agar and to remove a thread-like specimen immediately afterwards (Fig. 4). This is accomplished by using the tool first like a stirrer and then like a syringe. It needs, of course, some practice before the latter procedure can be performed perfectly single-handed.

When normal centrifuge tubes are used, a strip of material is removed from the pellet with the help of a small stainless steel spatula by pushing it below and across the pellet.

A large drop of a 0·2 to 5% solution of uranyl acetate in RK buffer is placed on a microscope glass slide beforehand. The specimen strip is laid into this drop. With the help of a stiff razor blade, a 1 mm wide band is cut out of the centre of the strip. This band is cut to cubes measuring at least $0·5 \times 0·5 \times 0·5$ mm and $1 \times 1 \times 1$ mm at most. The specimens are now ready for post-fixation (Fig. 5).

If the micromethod is used, the aspirated thread-like specimen is

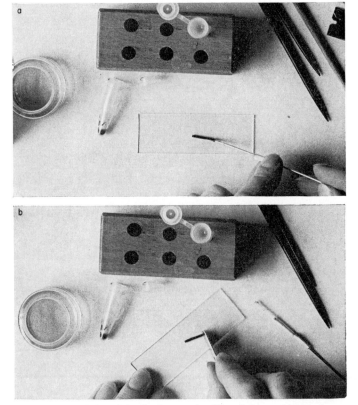

FIG. 6. (a) With the help of the "micromethod" tool shown in Fig. 4, a specimen-containing cylinder is pressed into a drop of uranyl acetate in RK buffer on a microscopic slide. The polyethylene vial to the left contains the remains of the pellet/Agar mixture. The glass container holds the uranyl acetate solution. (b) Cutting into small slices with the help of a stiff razor-blade.

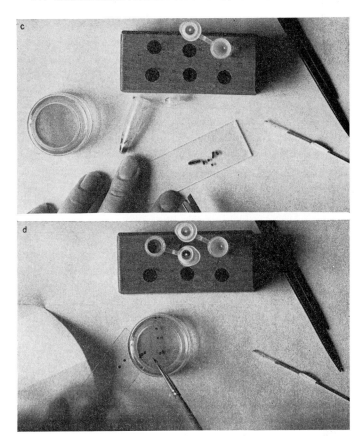

FIG. 6. (c) A few moments later, sufficient slices have been produced. (d) Slices are transferred into the uranyl acetate solution with the help of a small brush.

squeezed into the uranyl acetate drop and sliced there. Slices should not be thicker than 1 mm (Fig. 6).

(d) *Post-fixation*. Gelification of the bacterial nucleoplasm takes place during pre- and main-fixation. The stabilized nucleoplasm is fixed by the use of uranyl acetate, and this preparation step is called post-fixation.

The uranyl acetate used needs to be free of radioactivity. This radiation-depleted chemical can be handled with confidence.

For post-fixational purposes, a 0·2 to 5% solution of uranyl acetate in RK buffer is used. The stock solution must be stored in a ground-glass stoppered dark bottle, or better, because the contents can be checked easier, in a transparent bottle completely wrapped in aluminium foil. The reason for this is that uranyl acetate solutions are very light sensitive. The bottle must not be stored in a refrigerator. The stopper and its seat should always be kept clean, since crystallized residues tend to block the stopper.

For each specimen, a few millilitres of post-fixative are poured into a small glass beaker with a lid. The specimen cubes are transferred into the fluid from the microscope slide with the help of a very small brush (Fig. 5). This is the most suitable tool for all manipulations in the course of embedding procedures. Small objects stick excellently to the hairs and the fine tip of the wet brush permits extremely sensitive manipulations. On the other hand, the wet brush can easily be dried, if necessary, simply by touching filter paper, so that unwanted transfer of fluid can be prevented.

The duration of post-fixation is 2 h. The covered glass beakers are shaken from time to time or, even better, put on a rotary shaker.

The use of a rotary shaker is recommended, starting at post-fixation.

(e) *Dehydration.* Water in the specimens has to be completely replaced by an organic solvent, which, in turn, is miscible with the chemical used for embedding. Depending on the chemical properties of the embedding agent ethanol or acetone are used for dehydration. Out of a large variety of embedding substances, polyesters and epoxy resins fulfill all critical requirements of microbiological TEM techniques. Both are miscible with acetone. Ethanol therefore will not be discussed here.

The schedule for dehydration with acetone is as follows:

30%	15 min
50%	15 min
70%	15 min
90%	15 min
100%	2 times 30 min

There is a low percentage of water in pure acetone as it is supplied. It ranges around only 0·2% but cannot be tolerated under any circumstances. Many a failure in conventional microtomy of bacteria has been caused by badly dried acetone. This cannot be overemphasized. The most effective and most reliable drying agent is molecular sieve. The 3 A quality is recommended for the absorption of water. An ideal drying agent would indicate by a colour change when its drying capacity was exhausted. Molecular sieve does not have this quality. The best way out of this relatively harmless dilemma is the addition of sufficient molecular sieve into the original stock bottle, which is discarded after emptying and the regular replacement of used molecular sieve in those stock bottles, which are refilled. Ten grams of molecular sieve should be added per litre of EM grade acetone. The date of addition of molecular sieve is noted on the label. The frequency of replacement obviously depends on consumption and should be based on the premise of better too often than not often enough.

For dilution of acetone, twice-distilled water is used. 100 to 200 ml bottles are recommended for the schedule series. The appropriate amount of molecular sieve must be added to the 100% bottle. All bottles must have ground-glass stoppers.

It takes about 24 h before molecular sieve has quantitatively absorbed the residual water.

Transport of specimens again is performed with a small brush. Of each acetone solution a few millilitres are placed in a fresh glass container, moments before it is actually needed. Never try to remove one solution with a fine pipette and to pour in the next one. This inevitably leads to badly embedded bacteria. The specimen supporting brush is lightly pressed against filter paper whilst observing it carefully. The aim of this manipulation is to remove as much of the previous solution as possible, carefully avoiding even superficial drying of the specimen, which should always appear moist and glossy. Drying takes place fastest with the higher concentrations of acetone and, therefore, is most critical when the 100% stage is reached. If a specimen becomes greyish here, it must be discarded.

Breathing into the 100% acetone must be avoided.

The specimen-containing glass vessels should be shaken about every minute or put on to a rotary shaker.

(f) *Impregnation.* In this phase of preparation, acetone is replaced by a polymerizable chemical.

(i) *Epoxy resins.* In this class of substances, Epon 812 proves to be an outstanding medium (Glauert, 1974; Luft, 1961; Luft, 1973), provided the so-called "weight per epoxide equivalent" is known. This figure ranges from 150 to 170, according to one of the manufacturers (Ladd, 1975).

The replacement of acetone is performed in four steps with reference to the following schedule.

I	3 parts water-free acetone : 1 part Epon 812 mixture	30 min
II	1 part water-free acetone : 1 part Epon 812 mixture	30 min
III	1 part water-free acetone : 3 parts Epon 812 mixture	30 min
IV	Epon 812 mixture	2 times 30 min

Mixtures I to III are prepared by shaking in a stoppered measuring jar, while mixture IV is stirred by hand or with the help of a commercial stirrer.

The ready for use mixture of Epon 812 is composed of two solutions, of which one (solution A), contains Epon 812 and dodecenyl succinic anhydride (DDSA), and the other one (solution B), Epon 812 and nadic methyl anhydride (NMA). To these components, an accelerator is added, {tri(dimethyl amino methyl)phenol}, (DMP 30).

The weight of the anhydrides DDSA and NMA, (W_A) is calculated as follows:

$$W_A = 0.7 \frac{W_R M_A}{W_E}$$

W_R weight of Epon 812 sample,
M_A molecular weight of DDSA or NMA,
W_E "weight per epoxide equivalent",
0·7   ratio of the molecular weights of anhydride and epoxy resin.

The molecular weight of DDSA is 266, the molecular weight of NMA is 178.

If, for instance, W_E is 159 and 100 g Epon 812 were weighed in, the calculation becomes:

$$W_A \text{ (DDSA)} = 0.7 \frac{100 \times 266}{159} = 117.1 \text{ g}$$

$$W_A \text{ (NMA)} = 0.7 \frac{100 \times 178}{159} = 78.4 \text{ g}$$

117 g of DDSA are added to 100 g of Epon 812. This is mixture A. 78 g of NMA are added to 100 g of Epon 812. This is mixture B.

The use of mixture A alone would lead to soft blocks, the use of mixture B alone to hard blocks. The hardness of blocks, therefore, needs to be adjusted by the ratio mixture A : mixture B. In practice, for microbiological TEM, a mixture of 1 part mixture A and 9 parts of mixture B is found to be the most suitable.

Mixtures A and B can be stored at 4°C for many months. The original chemicals, however, must be kept at −20°C.

Before starting work with Epon 812 components, they must be kept at room temperature for several hours. If this measure is not observed, condensed water will be formed in the opened cold bottles, and their contents are then ruined. Mixtures A and B should be stored in Erlenmeyer flasks and the date of production noted upon them. They should be wrapped in aluminium foil, so that condensed water is strictly confined to this envelope after taking the flasks out of the refrigerator.

The final Epon 812 mixture is supplied with 0·14% (v/w) of the accelerator DMP 30, which should be stored and handled according to the precautions given in the last paragraph.

If, for instance, a total of 10 g of mixture A and B is prepared, 0·14 ml of DMP 30 must be added with the help of a 1 ml pipette.

The final Epon 812 mixture is stirred by hand for 20 min (!) and this must be done in such a way that as few air bubbles as possible are drawn into the solution. If a motor-driven stirrer is used, and this is recommended, the stirring time can be reduced to 10 min.

Rubber gloves should be worn during the manipulations because DMP 30 possesses skin irritant properties.

Transfer of specimens from the last change of acetone to impregnation step I and from here to II, III and IV, is again performed using a small brush. The brush drying procedure requires more time at this stage because of the high viscosity of the epoxy resin mixtures.

The use of a rotary shaker is strongly recommended during impregnation.

It takes 2·5 h to complete the impregnation. At raised room temperatures, it will be observed that the epoxy resin mixture in the glass containers starts to become more viscous as a result of polymerization. If this is the case, the preparation of a fresh portion of the final mixture, large enough to complete the impregnation procedure, is strongly recommended.

Dry (!) gelatin capsules are filled with Epon 812 mixture and one black coloured specimen is placed in each of them.

(ii) *Polyesters*. For a long time, microtomy of bacteria and the use of a polyester as impregnation medium have been synonymous, because the pioneer work in this field involved the first and very successful use of polyesters as embedding material (Kellenberger *et al.*, 1956). Polyesters have now lost their leading position in this field, but still are "primus inter pares".

Vestopal W is the best known polyester for impregnation. W stands for "weich", which means soft. Vestopal H, with H for "hart" or hard, is also manufactured but is useless for our purpose.

A Vestopal W kit comprises resin, "initiator" and "activator". The initiator is butyl perbenzoate, the activator is cobalt naphthenate. All these substances have to be stored at 4°C. The precautions mentioned above for storage of Epon 812 have to be carefully born in mind here also.

The impregnation schedule for Vestopal W is as follows:

I	3 parts water-free acetone : 1 part Vestopal W	30 min
II	1 part water-free acetone : 1 part Vestopal W	30 min
III	1 part water-free acetone : 3 parts Vestopal W	45 min
IV	Final Vestopal W mixture	2 times 45 min

Mixtures I to III are prepared by shaking in a stoppered measuring jar. Mixture IV, however, needs to be very carefully stirred. Stirring must be performed in two steps. First 1% (v/v) initiator is added to the resin and immediately mixed for about 1 min. Then 1% (v/v) activator is added and also mixed for about 1 min. Then the complete mixture is stirred carefully by hand for 20 min or with the help of a stirrer for 15 min.

Unused and clean pipettes are used for the addition of initiator and activator. Vestopal W and initiator are stable for about 6 months, if refrigerated. The activator, however, is a very sensitive chemical. It has to be rejected as soon as traces of precipitate can be seen in the purple fluid. The date of delivery should be noted on each container, as already mentioned previously for other chemicals.

The use of a rotary shaker is highly recommended for the Vestopal W impregnation steps. If none is at hand, the specimen-containing glass beakers must be tilted almost continuously. Steps III and IV are far from easy for the laboratory assistant under these circumstances. Many failures observed when Vestopal W is used to embed bacteria, can undoubtedly be traced to insufficiently dried acetone and careless manipulation of step IV.

Sufficient volume of mixture IV should be provided, so that both step IV and the filling of gelatin capsules can be accomplished. Intolerable untimely polymerization should not occur.

(g) *Polymerization.* Immediately after the dry gelatin capsules have been charged with specimens, they are placed in the oven. It is not necessary to wait until all specimens sinked down to the tips of the capsules. As soon as heat is supplied, turbulent flow develops in the resin, and specimens move up and down. After completion of polymerization, the specimens should rest at the bottoms of their capsules. If they do not, probably something has gone wrong with the impregnation.

Polymerization is a two-step process. The first step takes 12 h at 45°C and the second one 48 to 72 h at 65°C. Prolonged sojourn time at the higher level does no harm to the specimens. The resulting unit of polymerized Epon 812 or Vestopal W and enclosed specimen is called a block.

Often the quality of blocks improves after prolonged storage at room temperature or at about 50°C.

(h) *Sectioning.* Contrary to histology, large sections having lengths of sides in the order of magnitude of 1 mm are not needed for bacteriological TEM. This considerably eases microtomy work.

After removal of the gelatin capsule from the hardened block, mechanically or with the help of water, the specimen can clearly be identified with the naked eye. But not only this. If relatively few bacteria have been embedded with agar, the specimen looks piebald. The density of polymerized resin is almost equal to that of empty agar. Thus the black zones in a piebald specimen are the bacteria-bearing areas. In many cases, however, the specimen-containing agar cube shows a relatively homogeneous distribution of bacteria and is uniformly black.

The general rules of microtomy are pertinent to all fields of TEM. They should be studied by attendance at specialized courses or, less commendably, by self-training with the help of the manual of the microtome at hand.

Mechanical trimming of the block is the ideal method (Reid, 1974). In the present case this is performed in such a way that either the whole resulting trapezoidal surface of the pyramid looks black or that a black spot of a piebald specimen lies in the centre of the first cut. The medium width of the first cut needs not be larger than 0·3 mm, its depth about 0·1 mm.

Provided bacteria are randomly distributed in the specimen as is often the case, and supposing that 30% of the section is covered with cocci, then roughly 12,000 bacteria will be seen on a symmetrical trapezoidal surface measuring $360 \times 245 \times 100 \ \mu$m.

Epon 812 blocks cannot be perfectly sectioned with glass knives. Interactions occur between the surface of the curved knife edge and the cross-linked epoxy resin, leading to rupture of the sections. Diamond knives have to be used for sectioning Epon 812 blocks. Though diamond knives are costly purchases, they are worthwhile investments. In the hands of experienced and skilful assistants they last for many years. They can, of course, be used for Vestopal W blocks too, but here glass knives are adequate in most cases. If serial sections are necessary, for instance for the reconstruction of bacterial fine structures, a diamond knife considerably eases the work.

Very thin sections can be more easily made when the pyramid is small. Thus embedded bacteria may be considered as ideal objects for TEM at relatively high resolution, e.g. 2 nm. Sections can be made thin enough to apply darkfield electron microscopy successfully (Lickfeld, unpublished results).

Section thickness can roughly be estimated from the interference colours of floating sections. All greyish shades indicate section thicknesses between about 30 and 60 nm. Silver colours range from about 60 to 90 nm. Based on our own experiments with diamond knives, Epon 812 blocks, LKB microtomes, and measurements performed with a light optical interference microscope, it can be said that the nominal section thickness as shown on the built-in meter of the microtome, must be multiplied by a factor of about 1·4 to obtain the real thickness. Thus a 30 nm reading is to be correlated with a 42 nm slice thickness ± some 10%. If the surface of the pyramid on the block is made small enough, an easy task with mechanical trimming, extremely thin sections can be produced. They may be so thin that no interference colour emerges from them. This may happen, in a cool microtomy laboratory, with meter readings of between 5 and 10 nm. Real thickness may thus range between 7 and 14 nm. The discrepancy between nominal and real section thickness is partly due to compression of sections during the sectioning process. This leads to a decrease in section depth and a proportional increase in thickness. The latter ranges around 15 ± 5% for Epon 812 and less than this for Vestopal W. Compression inevitably leads to deformation of bacterial fine structures and must be taken into account when quantitative measurements of whole cells, or of any fine structure in them, are being made. Precursors of bacteriophages are well suited as

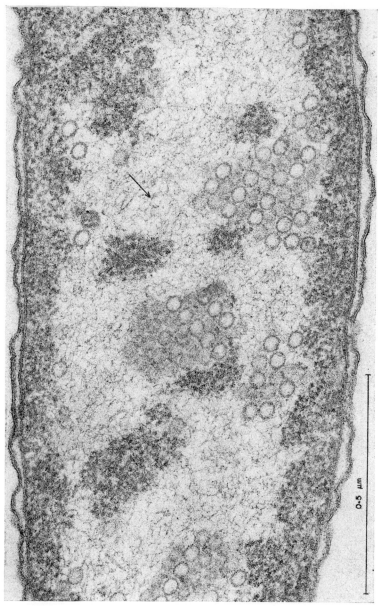

Fig. 7. Partial view of a median slice of *Escherichia coli* B, containing an A⁻ mutant of bacteriophage λ. Perfect preservation of all cell-related fine structures and of bacteriophage precursors (pλ), which show paracrystalline arrangements. *In vivo*, these particles are isometric. Due to compression of section, they show an elliptic deformation, from which the orientation of the cutting knife edge in regard to the specimen can be deduced. The arrow gives the direction perpendicular to the knife edge and parallel to the small axis of the deformed particles. Epon 812 embedding and pre-fixation with 0·2% (w/v) osmium tetroxide, final concentration.

objects to provide insight into the order of magnitude of compression (Lickfeld, 1975c). Roundish particles are deformed to ellipsoid shapes, and the short axis of particles of this type is always orientated perpendicularly to the knife cutting edge (Fig. 7). The ratio short axis : long axis, measured on a large number of particles, gives a compression figure from which the percentage of compression can be evaluated. Cocci are another, equally well suitable model for this evaluation. Before one tries to give absolute figures for the size of embedded specimens, shrinkage of specimens during the embedding procedure must be taken into account. Volume shrinkage is distinguishable from linear shrinkage and may amount to 30% (Hayat, 1970).

Since sections normally contain a very large number of cells in bacteriological TEM, it is not of great importance when sections are "fished from above" (i.e. picked up from the water surface by lowering a specimen grid down on to them), that they are picked up in such a way that location and direction of a series of slices is not predictable. This rough fishing procedure has two undesired disadvantages, however. Firstly, a relatively high proportion of the sectioned bacteria are hidden by the bars of the specimen grid. Secondly, folds in the sections arise from compression, which, in turn, hide bacteria and give rise to secondary effects of contamination as soon as staining for contrast enhancement is performed. "Fishing from below" is a better technique and it can rarely be avoided when serial sections are being prepared, each one of which must be seen in the microscope. Serial sections should always be fished on slotted specimen grids with slots long enough to receive a whole (about 1 mm long) series of sections. Single slot specimen grids are equally well suited to this purpose.

(i) *Contrast enhancement.* Sections showing silver or even gold colours in the knife trough, in general may not need staining with heavy metal salts to enhance image contrast. Thinner sections however must be post-stained.

Contrast, of course, may be enhanced by use of a lower acceleration voltage in the microscope but this leads to more specimen damage because beam current has to be increased to ensure sufficient screen brightness unless an image intensifier is used.

Staining should be performed with lead citrate (Reynolds, 1963). The following solution is prepared.

Carbonate free (!) lead nitrate ($Pb(NO_3)_2$)   1·33 g
Sodium citrate ($C_6H_5Na_3O_7.2H_2O$)   1·76 g
     in 30 ml boiled distilled water

The solution is shaken for 30 min. During this time, lead nitrate is converted into lead citrate. Then 8 ml of a 1N solution of sodium hydroxide

(NaOH), 3·99 g in 100 ml boiled distilled water, are added. The milky solution clears and boiled distilled water is then added to a total volume of 50 ml. The staining solution is ready for use. It is stable in a ground-glass stoppered bottle at room temperature for months, but should be discarded as soon as precipitates are observed.

Immediately before post-staining is performed, a row of drops of the staining solution is placed in one half of a plastic petri dish, a row of cold boiled distilled water drops in the other half. The charged specimen grids are laid upside-down on the stain solution drops with the help of clean, gilded and curved forceps. The reaction time must not exceed 20 sec, since otherwise precipitate spots will be formed consisting of very small electron opaque points with diameters of about 3 nm. If they form in the cytoplasmic area, they seriously interfere with genuine fine structures. After staining, the specimen grids are again held firmly with gilded forceps and carefully rinsed three or four times in the water drops.

Water remaining between the tips of the forceps is removed with filter paper before the specimen grids are laid down on filter paper for drying.

Forceps with tips fine enough to make them suitable for electron microscope work are made of non-stainless steel. Interactions between this steel and common staining fluids take place in such a way that precipitates are formed on the specimen grids. This annoying effect can be eliminated by the use of gilded forceps.

Breathing into the loaded Petri dish must be carefully avoided in order to prevent the formation of minute precipitates of lead hydroxide in the staining solution and on the sections.

Artifacts may arise if post-staining is prolonged (Fig. 8). Longer periods of staining may be satisfactorily carried out at low temperatures.

2. Cryomicrotomy

IMPORTANT NOTE: Unlike chemical fixation, rapid freezing does not of course kill bacterial cells. The processes of cryomicrotomy and freeze-etching necessarily involve the handling of living bacteria and are such as to create a serious contamination hazard. Adequate precautions must be observed at all times to prevent the spread of living cells by the use of a rigorous disinfection routine during these operations whether the organisms are known to be pathogenic or not. Microtomists who are not microbiologists should seek guidance from their microbiological colleagues. Useful information is to be found in the Chapter on Safety in the Microbiological Laboratory by Darlow in Volume 1 of this Series.

Cryomicrotomy is only in its infancy. The aim of cryomicrotomy is the study of thin sections of chemically unfixed bacteria in the transmission electron microscope. Attempts have been successfully undertaken to use

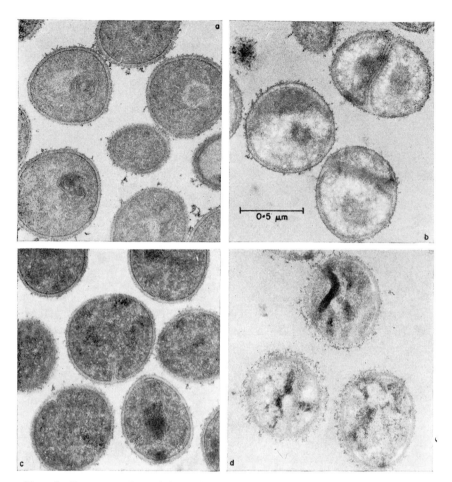

Fig. 8. Demonstration of lead citrate post-staining artifacts. In this figure, micrographs (a) to (c) show parts of Epon 812 sections of equal thickness. (a) For comparison, this section remained unstained. (b) Staining with lead citrate was performed for 10 min at room temperature (22°C). The interior of the *Staphylococcus aureus* cells looks spongy due to unknown reactions. (c) Contrary to (b), post-staining was performed for 10 min at 4°C, "in the cold". Electron opacity has increased considerably as compared to (a). In the cytoplasm, minute processes comparable to those, which led to artifacts shown in (b), can be discerned. In practice, reaction time is therefore drastically reduced by a factor of about 30. (d) Part of a Vestopal W section, post-stained with lead citrate for 10 min at room temperature. The upper cell exhibits a typical artifact, which is the result of a shrinkage process within the cell. This effect can be eliminated by post-staining at 4°C.

wet specimens in an electron microscope (Parsons *et al.*, 1972), but they are restricted to intact, unsectioned specimens because any attempt to section genuine biological material is *a priori* condemned to failure because of the severe physical damage brought about by the shearing forces of any knife. Specimens must be made stiff in order to reduce artifacts caused by sectioning to a minimum. As the term "cryo" implies, specimens are made rigid for sectioning with the help of cold. Specimens for sectioning must be prepared by very rapid freezing. If freezing is fast enough, the formation of ice crystals can be reduced so effectively that the seize of ice crystals is below or just at the limit of the resolving power of the electron microscope. This condition is called vitrification. Discussion continues as to whether vitrification in the true sense of the word really exists. But this is a problem which is far beyond the scope of this Chapter.

Basically, cryomicrotomy consists of four preparational steps, freezing, sectioning, fishing of sections, transfer of sections into the microscope. Each step presents technical problems. On the other hand, cryomicrotomy promises to become one of the most powerful techniques in biology in general and in bacteriology specifically because no chemical fixation, no dehydration with the help of organic solvents and no embedding in monomers and polymers is involved. Admittedly, cryomicrotomy is still in its early stages, but its potential presents a challenge to improve the methods used.

There are a few microtomes on the market, which are suitable for cryomicrotomy with the help of commercially produced cryo-attachments. The "FTS"-attachment has been constructed for the Sorvall microtomes "MT-2" and "MT2-B". The LKB microtomes, "Ultratome I and III", can be completed with a so-called "cryokit", and the Reichert microtomes "Om U2 and 3" with the cryo-attachment "FC 2". To become acquainted with the fairly complicated systems, a high level of competence with the "naked" microtome is essential. This craftsmanship is best earned by attendance at university or factory courses and by subsequent experience. The starting-place for cryomicrotomy is a thorough knowledge of the cryo-attachment available. Cryomicrotomy is a full-time activity for skilful and experienced assistants or scientists.

Electronic temperature control between $0°C$ and $-150°C$, with an accuracy of 2%, is provided in the Sorvall cryo-microtome system. The modified LKB microtome allows one to choose knife temperatures between $-70°C$ and $-140°C$, and specimen temperatures between $-70°C$ and $-170°C$. Temperatures are kept constant electronically within $\pm 0\cdot2°C$. In the modified Reichert microtome, the temperature of knife and specimen can be preselected between $+40°C$ and $-160°C$. They are constant to better than $\pm 0\cdot2°C$. In practice, temperatures above $-80°C$ must be avoided to prevent recrystallization of the frozen specimen. Experience is necessary to determine the combination of temperatures for knife and specimen that are optimum for preservation of fine structures. It is important to recognize that so far no-one knows what "preservation" really means because nobody so far has observed *in vivo* fine structures in the transmission electron microscope. A crucial criterion is the appearance of membranes, that is of cell wall and cytoplasmic membrane, in suitably orientated bacteria. They should appear as continuous envelopes without any damage whatsoever.

(a) *Rapid freezing.* If possible, plate cultures are used for cryomicrotomy because liquid cultures have to be centrifuged before preparation without the use of a chemical fixative. If this is done, unpredictable modifications of the bacterial fine structures may occur as a result of oxygen deficiency. If liquid cultures must be used, centrifugation should be made as short as possible, just yielding enough of a spongy pellet.

With the help of a platinum loop, a small portion of the bacterial preparation is placed on the specimen holder of the cryomicrotome in such a way that a conical tip is formed. This unit is then dropped into cooled propane or Freon 22. This is a very critical step.

The smaller the specimen and the lower the thermal capacity of the specimen plus holder unit, the more probable it is that a state of vitrification will be obtained. The best approach is to use very small drops, as will be shown for spray-freeze etching, but as far as is foreseeable, this method cannot be used for cryomicrotomy.

Propane is the best suited coolant for rapid freezing (Bachmann and Schmitt-Fumian, 1973). Pure propane, however, should not be used, but instead propane of technical quality. The melting point of propane is $-189 \cdot 69°C$. It is cooled down to this temperature by means of boiling liquid nitrogen at $-195 \cdot 8°C$. The device used for cooling down propane is either a commercial model or a copy of such a unit (Fig. 10b). There is a temperature gradient between the liquid nitrogen level in the Dewar container and the propane level, which, usually, prevents the propane from being frozen. But if freezing occurs, the brief introduction of a clean stainless steel rod will melt the frozen propane. Liquefaction of propane is performed by carefully blowing a stream of gaseous propane close to the bottom of the hole in the insert of the Dewar container through fine copper tubing. Theoretically, mixtures of gaseous propane and nitrogen may be explosive, but in practice this has never been experienced, presumably because of the rapid diffusion of both components into the surroundings. Nevertheless, safety goggles should be worn during these manipulations.

In principle, manipulations are the same as those just described, if Freon

22 is used. The melting point of Freon 22 is $-160°C$. Due to the large difference between the boiling point of liquid nitrogen and the melting point of Freon 22, namely $35·8°C$, Freon 22 freezes in the insert of the Dewar container. Before use, it is melted as described in the last paragraph.

The aim of the application of Freon in connexion with any rapid freezing experiments in biology is the suppression of the Leidenfrost phenomenon, which may retard the freezing velocity. The physical–chemical properties of Freon predispose it to eliminate this hampering phenomenon, unlike propane. Nevertheless, experimentation with propane has emphasized its predictable excellent freezing properties.

After rapid freezing, specimens may be stored in liquid nitrogen. Transfer must be made as rapidly as possible by means of cooled (!) forceps. Before use, these are dipped into liquid nitrogen until boiling ceases. The handle of the forceps should be wrapped with plaster tape of high thermal resistance to enable the experimenter to use them without getting "burned" from ascending cold. The storage system in the respective Dewar container should be similar to commercially available ones (e.g. Balzers, Liechtenstein).

(b) *Sectioning*. Another extremely critical step of cryomicrotomy is the transfer of the rapidly frozen specimen from the freezing bath or from the storage bath to the cooled specimen holder of the cryomicrotome. Again pre-cooled forceps are used and attachment of the mounted specimen must be performed so rapidly that even the slightest sign of thawing around the specimen is prevented. Otherwise the specimen should be discarded. This procedure should be practiced with the help of an empty specimen holder.

Another difficulty encountered here, is the formation of hoar-frost. Freon 22 is used as an anti-hoar-frost agent. A small brush is used to dab the surface of the endangered area. It looks thawed afterwards but is at the preset low temperature of the clamping system of the cryomicrotome a few seconds after application.

A method used so far to pick up frozen sections is similar to that used for conventional microtomy. They are fished from the knife trough, which is filled with a mixture of distilled water and dimethyl sulfoxide. This no doubt is a rough compromise, intolerable as soon as unfixed bacteria are used.

For pure cryomicrotomy, "dry" thin sections have to be made, that is they have to be collected directly from the cooled and dry knife edge.

Sectioning of frozen bacteria proves to be very easy. With the LKB cryomicrotome, for instance, pre-set nominal thicknesses of 5 to 10 nm are practical when $45°$ glass knives of appropriate quality are used. Even thinner sections are easily obtained with a diamond knife. In either case, shearing forces cause arched thin sections, contrary to conventional microtomy where, due to the elasticity of the polymers, thin sections flatten after being cut.

Fishing of "dry" sections is a crucial step, which needs considerable practice. A pre-cooled specimen grid must be used for fishing, and the series of sections has to be transferred into the column of the microscope in such a way that the temperature of sections and grid do not exceed $-85°C$. At the same time, contamination from condensing water vapour must not occur. Recently a device was demonstrated, which seems to indicate a way out of this problem. Although this JEOL "Cryo-chain" is specifically designed for the JEM 100 C, the principle can be applied to any transmission electron microscope or at least to those with immersion objective lens.

If one succeeds in transferring the frozen thin sections into the microscope, the last and inescapable preparational step, freeze-drying, takes place in the high vacuum of the instrument. The specimen holder must remain cold during freeze-drying as it is secured with the "Cryo-chain" device. If this cannot be achieved, freeze-drying of the frozen thin sections, mounted on specimen grids, must be performed in a commercial freeze-drying apparatus. The time necessary for complete dehydration is determined experimentally, because the sublimation rate of ice depends critically on its temperature. Drying should be complete after between 10 and 100 min.

After drying of thin sections, a lock device must be used to ensure absolutely moisture-free transport from the freeze-drying unit into the microscope column. At present, specific locks must be designed and built by the operator. Top-entry type objective lenses (e.g. "EM 10" or "Elmiskop 101 and 102") require more complicated lock devices than side-entry objective lenses (e.g. "EM 201, 300, 301").

(c) *Use of stains.* Due to lack of heavy metals, cryo-sectioned specimens exhibit very low electron opacity which is exaggerated, since section thickness is very small. So far a kind of negative-staining has been used to enhance specimen contrast. But this was only possible because no freeze-drying of slices was performed. Pre-fixed bacteria were negatively stained after lifting from the collection bath of the microtome.

Pure cryomicrotomy excludes the use of stains. All efforts to gain an insight into the, admittedly fragmented, *in vivo* fine structures are ruined by the addition of electron-scattering stains. Electron optical measures must be taken instead to obtain a high enough image contrast. Contrast-stop (Dupouy, 1968) and darkfield TEM (Dubochet, 1973) are the methods of choice. Much experimentation is necessary before the full potential of such techniques can be determined.

C. Freeze-etching

1. *Conventional freeze-etching*

A serious disadvantage of any thin sectioning technique is the incapacity of the method to demonstrate the fine structure of surface aspects of bacterial membranes. A membrane running parallel to the surface of a section may either be obscured by adhering cytoplasm or it has, if separated, too low an electron opacity to be visible or it may be torn out from the section during sectioning, if it was too close to one of the surfaces. Freeze-etching is the only known method to show quasi *in vivo* aspects of biological specimens. It is particularly suitable for displaying the structure of surfaces (Bullivant, 1973).

Several freeze-etching devices are available or have been described recently (Koehler, 1973). They all incorporate fairly complicated mechanics and electronics. The crucial part of any freeze-etching device is the specimen holder in the bell jar. Its temperature is kept extremely constant electronically. The temperature range extends from room temperature to about $-150°C$, the $-100°C$ point being the most important in general. The actual temperature of the specimen holder when the meter reads $-100°C$ is often found to deviate from the indicated figure. Careful calibration using a microthermo-couple is essential if good results are to be obtained.

As a preliminary check, a drop of acetone is placed on the specimen holder at $-150°C$. After the final vacuum is reached in the bell jar, the temperature control is set back to about $-10°C$. The frozen acetone is continuously observed. Read the meter the moment the acetone starts to melt indicating that a true temperature of $-95°C$ has been reached.

(a) *Cultivation of bacteria for freeze-etching.* What has been said about cultivation of bacteria for TEM in Section III.B, 1 (a) applies likewise to freeze-etching.

(b) *Rapid freezing.* For rapid freezing of biological specimens, the term "physical fixation" is used. For physical fixation of bacteria, antifreeze agents should be avoided (Fig. 11), since they must be considered as unphysiological. If an antifreeze agent proves to be necessary, the type of antifreeze agent, its concentration and its reaction time need to be individually tested for the particular bacterium (Hagen, 1971; Nash, 1966). So far in most cases glycerol has been used as an antifreeze agent for bacteria at a final concentration of about 20% (v/v) (Nanninga, 1973).

Liquid cultures are the initial material for freeze-etching. A very small volume of centrifuged material is placed on a metal support, and both undergo rapid freezing simultaneously.

For transfer of bacteria to the support, a modified Pasteur pipette is used.

The original fine tip of such a pipette is pulled out by hand over a small burner in such a way that a fine tube with an outer diameter of about 0·05 mm results. This thinned section is broken and the pipette is ready for use.

Commercially made gold-nickel specimen supports are the most suitable due to their high conductivity and high mechanical strength. Either the type with a central disc or the type with a central ring is used. The resistance of the specimen to shearing forces is higher for the ring type but its use does not eliminate the danger of tearing out the specimen. Discs should be scratched to increase the adhesion of the specimen.

Fig. 9. With the help of a Pasteur pipette, viscous bacterial culture is filled into a specimen support of a freeze-etch apparatus. In (a) the procedure has just started, in (b) it was just finished. Here, the convex surface of the drop can easily be discerned.

A 1·5 ml portion of the liquid culture is pelleted in a table-top centrifuge at about 8000 *g* for only 3 min. The pellet needs to be fairly viscous but pipettable. While the specimen support is held with curved pointed forceps (Fig. 9), an appropriate portion of the pellet is placed on the disc or into the ring of the specimen support. The volume of the drop will be about 0·3 or 0·7 mm^3, respectively. Less than half a sphere of the specimen should be on the top of the support in each case.

The loaded specimen support is dropped (!), not thrown, into Freon 22 or, even better, Freon 13, which is cooled down to its respective melting point ($-160°C$ or $-181°C$, respectively), in the same way as described in Section III.B, 2 (a) (Fig. 10a). Usually a series of specimen supports is prepared successively. The support is stored temporarily in liquid nitrogen. Transfer from Freon 22/13 to liquid nitrogen has to be performed as quickly as possible.

(c) *Fracturing*. Loaded specimen supports are transferred to the specimen holder of the freeze-etching device in use. This again is done with pre-cooled forceps as quickly as possible. Liquid Freon is used to remove hoar-frost from the specimen holder.

For fracturing, a degreased razor blade is used. Regrinding of the razor blade is unnecessary and indeed leads to worse fracturing and more slicing. From experience, fracturing speed should be high. A relatively coarse feed of the knife arm is performed manually.

FIG. 10. In (a), the introduction of Freon 22 into a commercially made insert (Balzers) is shown, which is cooled by means of liquid nitrogen in a Dewar container. In (b), the introduction of propane is shown. The device shown here is part of a spray-freeze apparatus (see Fig. 14). The surface of each liquefied medium can clearly be discerned.

Our own experiments have shown that both the outer region of the specimen drop and its centre are satisfactorily frozen. This is in full agreement with theory (Vennrooij, 1975). Therefore about half of the frozen drop is removed by fracturing before sublimation is started.

(d) *Sublimation*. The aim of sublimation is to heighten the surface relief of the fracture face in the vertical axis. Fine structural details are better displayed in this way since their visibility depends on their contrast with their surroundings, which, in turn, depends on the existence of more or less vertical planes on which electron opaque material can be deposited by shadowing.

For sublimation, the vacuum should be better than 5×10^{-7} Torr. Sublimation time has to be determined experimentally. One minute will be a good start for bacteria. There is no good reason why sublimation should be omitted, even though sublimation starts the moment the last knife stroke has been made. Omission of sublimation effects a loss of information on exposed fine structures.

The knife and knife arm are utilized for sublimation. They are positioned over the specimen and perform as a cryopump. The fractured surface is protected against contamination at the same time. In bacteriology, contamination is particularly annoying since contamination particles have about the same size as genuine bacterial components, such as ribosomes.

(e) *Shadowing and coating.* A replica of the genuine fracture through the frozen drop is made in a two-step process of shadowing plus coating. In principle, carbon coating could be omitted, because its only function is to stiffen the product of shadowing to the extent that it can safely be manipulated during further preparational steps.

In my opinion, the term "replica" is a misnomer when shadowing is an integral part of "replica" production. A replica is a continuous layer, for instance of lacquer, following each spatial detail of a rough surface. It can easily be shown that shadowing is far from yielding a continuous layer. It leads rather to a holey cover as long as the thickness of the shadowed material remains relatively small, that is 2 to 3 nm. It must be kept thin, in order to be able to demonstrate fine structural details. A real and perfect replica represents the unbroken negative of a discarded positive as is the mould of a machine part.

Shadowing inevitably yields artifacts. They are in functional relation to depth of sublimation, thickness of the platinum carbon layer, orientation of fine structure to evaporation source, and angle of shadowing (Lickfeld *et al.*, 1972; Lickfeld *et al.*, 1975). In addition, what can be seen on "replicas" of this kind in the electron microscope are in fact pseudo fine structures.

The production of these artifacts, in addition, leads to the elimination of genuine mirror symmetry of very small structural details. This must be taken into consideration when double fractures are made (Mühlethaler, 1973).

The resolving power of "replicas" is limited by physical–chemical properties of the evaporated substances and by the thickness of the platinum carbon layer (Abermann, 1973). Holes, for instance, with a diameter of 5 nm, normally resolvable in an electron microscope with ease, remain invisible in "replicas" due to asymmetric deposition of material on the hole wall, which causes a partial closing of the hole.

The smaller the fine structural detail being studied the lower must be the angle of evaporation. In general, angles around 30° are common. For demonstration of single DNA threads, as they exist in the bacterial nucleoplasm, very small evaporation angles of about 5° to 10° are of help.

The contribution of the superstratified carbon layer to information about the underlying fine structures can be ignored. It lowers the image contrast but is absolutely essential to confer on to the replica sufficient mechanical stability. A layer thickness of 20 to 30 nm should be achieved. Only then will the specimen lose its tendency to roll up and break while floating on cleaning fluids.

(f) *Aftertreatment*. The cleaning schedule for replicas is as follows:

Sulphuric acid	overnight
Rinsing in distilled water	once
Eau de Javelle	2 to 4 h
Rinsing in distilled water	5 times, 30 min each

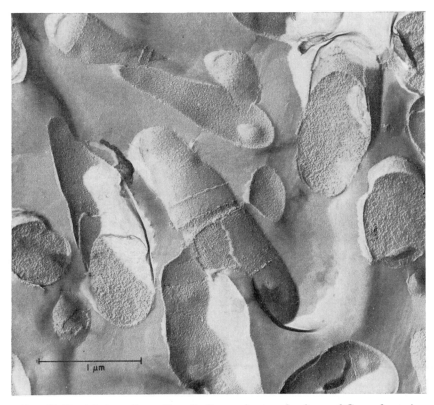

1 μm

Fig. 11. Partial view of a replica of a frozen-fractured culture of *Corynebacterium diphtheriae*, which was prepared without the use of an anti-freeze medium. Notice the freedom from segregation effects in the intercellular space. In this micrograph, "shadows" appear white. Compare with Fig. 12.

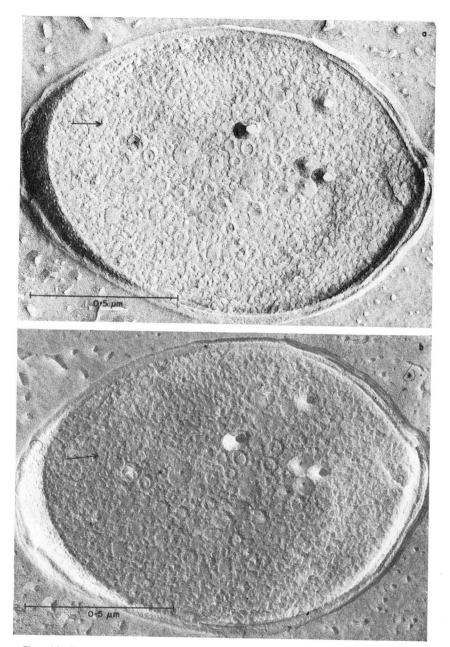

Fig. 12. Part of a replica of a spray-frozen sample of *Escherichia coli* B cells, containing wild type bacteriophage λ. (For the ring shaped particles, compare with Fig. 7.) (a) and (b) show the same cell, in (a) printed in the usual way, in (b), a double copy has been prepared. In (b), therefore, "shadows" are dark and the beam-lit areas are bright. Notice that both in (a) and in (b) the direction of shadowing is the same, as the arrow indicates.

Though the material adhering to the replica after removing it from the specimen holder is very spongy, cleansing proves to be very critical. A platinum loop is used for transfer manipulations. Bacterial debris is far better removed from replicas by use of sulphuric acid than by use of chromic acid.

(g) *Remarks on micrographs.* We are accustomed to dark shadows and bright sunlit areas. This often helps considerably in shape recognition in everyday life. By shadowing, as it is performed in the course of freeze-etch preparations, matter is deposited only at points exposed enough to "see" the source. "Shadows" are characterized by lack of matter.

Thus a fine structure related pattern of electron opaque and electron transparent areas arises. Prints of frozen-etched specimens are thus produced, on which "shadows" are white and beam-lit areas are dark. For the eye, this constitutes an embarrassing, unpsychological inversion and it severely hampers the interpretation of frozen-fractured surfaces. Convex and concave appearing details for example may easily appear reversed. This tricky phenomenon is multiplied by flaws in the presentation of light and shadow. It is therefore recommended that a negative contact copy of the plate or film be produced, from which prints can then be made (Fig. 12).

2. *Spray-freeze technique*

The most crucial part of any cryo-technique is rapid freezing. Freezing velocities of 400°C/sec can be obtained with specimen sizes such as are usually found in conventional bacteriological freeze-etching. Under such conditions, anti-freeze agents are unnecessary when bacteria are prepared in residues of nutrient medium, which may themselves function like an anti-freeze agent (Fig. 11). Without the addition of an unphysiological anti-freeze agent, even the slightest deviations from critical parameters, for instance size, will result in useless specimens.

Size does not play a predominant role as soon as specimens with diameters ≤ 50 μm are used (Bachmann and Schmitt-Fumian, 1973; Plattner *et al.*, 1973). The spray-freeze technique elegantly fulfills this condition. An appropriate commercial spray-freeze device has just become available. Earlier, home-built devices had to be used (Fig. 15).

Unlike harmless specimens (e.g. non-toxic macromolecules), bacteria must not be sprayed in a laboratory, whether they are pathogenic or not. Spraying must be performed in a separate room, which can easily be disinfected. In practice, there is no choice other than an incubator-like safety-device (Lickfeld *et al.*, 1976b). All manipulations up to the point where the specimen is ready to be fractured are carried out in the hermetically sealed safety-device (Fig. 13). Disinfection is performed from below through the receptacle hole of the safety-device.

Propane at its melting point, − 189·69°C, is used for rapid freezing. The propane container is an integral part of a copper rod that plunges into liquid nitrogen (Fig. 14). Gaseous technical (!) propane is carefully blown into the cooled-down copper rod receptacle, until the liquid propane level is a few millimetres below the flange surface. Then immediately the Dewar container/copper rod unit is coupled to the safety-device from below. At once a copper lid with central hole (Fig. 14), cooled in liquid nitrogen, is put on the copper rod. This is an anti-hoar frost measure. Ten of these lids should be kept in stock in liquid nitrogen. Spraying may start now.

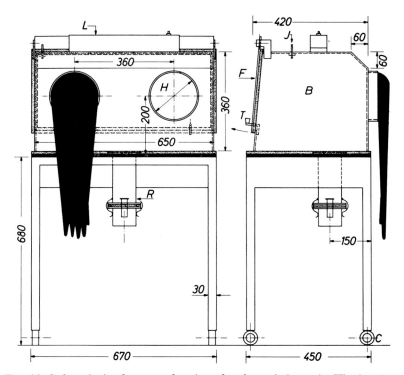

Fig. 13. Safety-device for spray-freezing of pathogenic bacteria. The incubator-like Plexiglass box B is hermetically sealed. It is illuminated by cold-light source L. All ancillary instruments are inserted through the opening in the back of the box, which is covered by the hinged flap F. The opening is sealed by rubber strips (dotted areas) and a pair of clamps T. For manipulations, two rubber gloves (one shown only) are at hand, which are fastened to appropriate holes H. Compressed air for the spray gun is introduced via J. The Dewar container and its accessories (see Fig. 14) is held by receptacle R, which consists of two parts, clamped together. Disinfection after completion of preparatory steps is done through opened and emptied receptacle R. All dimensions in mm. (By courtesy of *Microscopica Acta*; Lickfeld *et al.*, 1976a).

Fig. 14. The copper rod *a* with its hole in the centre of the flange is fixed in the Dewar container by ring *b* (see Fig. 10b). Anti-hoarfrost lid *c* is put on the flange of *a* in such a way that its 0·5 mm high ring projects into the hole. After completion of spray-freezing and for transport, cover *d* is put over the flange of *a*.

Thirty millilitres of liquid culture are distributed in four tubes and centrifuged in a table-top centrifuge at about 3000 *g* for 5 min. Each pellet is immediately resuspended in 1 ml broth.

The spray gun is loaded with an appropriate portion of the initial material. Spraying is performed in short bursts every 15 sec. Each burst heats the propane, and intervening periods are necessary to let the propane cool down again. Spraying is repeated until the propane looks milky.

At intervals hoar-frost covered lids are replaced by unused ones. As little hoar-frost as possible must fall into the specimen-containing propane. The more milky the spraying bath looks, provided this is not caused by ice crystals, the easier it will be to perform the following preparational steps.

When spraying is finished, a protective lid (Fig. 14) is put over the copper rod flange. The Dewar container is immediately removed from the safety-device, and the copper rod is taken out of the Dewar container with the help of cooled pliers. It is put into the appropriate borehole of the aluminium block in the spray-freeze accessory (Fig. 15).

Immediately after insertion, the specimen-containing propane in the hole of the copper rod is removed by pumping. The valve between flange and rotary pump needs to be carefully adjusted to provide very low pumping power, so that propane only is removed—and not the complete contents of the hole. What stays behind, is a white, dry and brittle substance. To this, about an equal volume of cold n-butylbenzene is added with the help of a modified Pasteur pipette with pulled, very fine tip. In most cases, one drop of n-butylbenzene is sufficient.

The consistency of the specimen/n-butylbenzene mixture must be pappy, so that it will adhere firmly to specimen supports. The mixture must be carefully homogenized by stirring, using a platinum wire, which is inserted into the tip of a copper cone, which, in turn, acts as a cold reservoir.

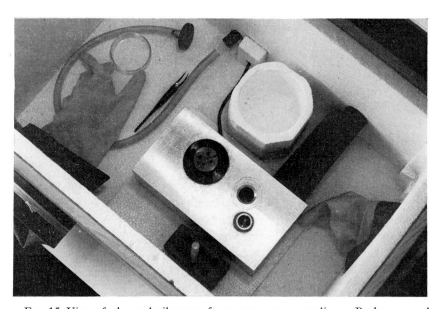

FIG. 15. View of a home-built spray-freeze accessory according to Bachmann and Schmitt-Fumian (1973). It consists of a wooden box with Plexiglass lid. The walls of the box are covered with 2 cm thick pieces of plastic foam to increase insulation. In the centre of the box, an aluminium block is visible. It is cooled with liquid nitrogen, which flows through spirally running channels. The temperature of the block is electronically kept at $-85°C$. The block has three boreholes, one of which is closed by a flange, which includes a thermo-couple for temperature control plus a specimen support clamping device. In the photograph, this is the uppermost part of the block. The borehole in the middle receives the specimen-containing copper rod (part a of Fig. 14). The third borehole contains a tube, which holds n-butylbenzene. In the upper left corner of the photograph, a plastic flange is visible. This is connected to a rotary pump via rubber tubing and valve. For manipulations, rubber gloves are used, fastened to holes in the walls.

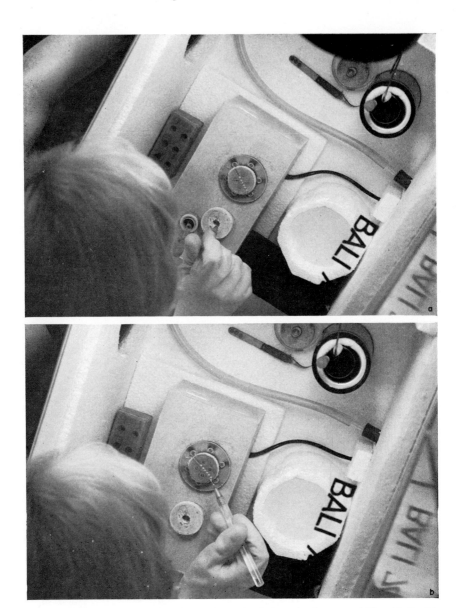

Fig. 16. Two photographs of actual manipulations in the box of Fig. 15. In (a), the spray-frozen specimen is mixed with cold n-butylbenzene. This is done with a cooled platinum wire, which is part of a copper cone. This instrument can better be seen in (b). In (b), commercially made gold-nickel specimen supports are loaded with appropriate portions of the soft specimen/n-butylbenzene mixture. The loaded specimen supports are dropped into liquid nitrogen. An appropriate Dewar container can be seen in the upper right corner of (a) and (b). The plastic container to the right is also filled with liquid nitrogen. This is used for cooling the stirring instrument wire plus copper cone every 15 sec.

This instrument (Fig. 16b) has a plastic handle and is cooled down in liquid nitrogen about every 15 sec (!) until boiling ceases.

The pappy substance (Fig. 16a) is loaded on to cooled specimen supports (Fig. 16b) which are dropped into liquid nitrogen.

All further preparative steps are according to the normal freeze-etching procedure, keeping in mind, that rules of disinfection are to be strictly observed after completion of the operations.

IV. ANCILLARY METHODS

A. Quantification of particles in bacteria

A method, which was originally developed to quantify from thin sectioned bacteria the number of bacteriophage particles per bacterium (Séchaud et al., 1959), is applicable to any type of uniform particles that may be occasionally observed, e.g. vesicles in certain lithotrophic bacteria.

Chemically fixed, embedded bacteria are randomly distributed in a block. After thin sectioning, sections of bacteria in all possible planes and directions are imaged on the fluorescent screen, from very faint polar slices and oblique sections, to impressive median sections. For quantitative measurement of particle numbers, at least 100 of the bacterial sections should be evaluated. Section thickness should be constant. One and the same thin section therefore is used throughout. Counting is performed directly from the screen. Each thin section is systematically scanned, carefully avoiding double counting. Since a technical assistant is necessary for notes during the operation he or she should also check that double counting does not occur. It is of utmost importance that all sectioned bacteria, even the smallest fragment of one, are taken into account.

The total number of particles counted is divided by the number of scanned sectioned bacteria. This result is multiplied by a factor, which depends on particle size and section thickness, and must be calculated according to the original publication.

For T-even bacteriophages and 40 nm thick sections, for instance, the factor is about 20. Thus if 1540 60 nm particles were found in 175 bacterial sections, the number of particles per cell would be about

$$\frac{1540 \times 20}{175} = 176.$$

Section thickness is measured with the help of light optical interference microscopy or directly in the transmission electron microscope (Edie and Karlsson, 1972).

It should be recognized that particle counts from thin sections inevitably yield figures which are too low, because polar sections of particles may fall out during sectioning or they have such a low optical density that they cannot be discerned, either on the screen or on micrographs.

B. Size measurements

Particle size measurements for size distribution are made on micrographs. Both the precise magnification of the microscope and that of the enlarger in the dark room must be known.

The microscope is calibrated with the help of a grating replica (see p. 173). Calibration is made immediately after taking the picture, for which the exact magnification has to be determined, or a series of pictures, which have been taken at a constant, but still unknown, magnification. All controls of the microscope are left as they are set. The grating replica is focused after insertion. Then a picture is taken.

If, for instance, a 2160 lines per millimetre grating replica is used, the magnification is

$$M = 2160 \left(\frac{d}{s}\right)$$

where d = distance between outermost grating replica lines on plate or film in mm and s = number of spaces between these grating replica lines. For example: $d = 78$ mm, $s = 7$, $M = 24,069$ times.

The calibration of a dark room enlarger is performed with the help of any test negative, which shows appropriate structures, preferably a grating.

C. Stereoscopy

In TEM, the only application of stereoscopy is to be found with frozen-fractured specimens, because here height differences of a replica are large enough to generate a spatial illusion when a conjugated pair of micrographs is viewed with a special double magnifier. Very often additional information on the relief of a frozen-etched specimen is extremely valuable for studying such method-related parameters as, for instance, the depth of sublimation. The apparent transformation of two flat micrographs into a single spatial impression is enhanced when prints of copied negatives are used, as described in Section III.C, 1 (g).

Taking stereoscopic pictures with a transmission electron microscope is performed simply by taking two pictures successively, with the specimen tilted $+6°$ and $-6°$, respectively.

In most cases, the tip of a special cartridge is tilted. Eucentric goniometers are very helpful for stereoscopy. With them, tilt angles can be chosen so large that statements concerning the spatial distribution of fine structures can be made even while observing the images on the fluorescent screen.

ACKNOWLEDGEMENTS

I am indebted to Eduard Kellenberger. Many a stimulus, received from him, eased the struggle for improvement of some specific methods in bacteriological TEM, which are part of this contribution. I am grateful to Beate Menge and Ursula Almert, whose perseverance played an important role, particularly in the domain of freeze-etching techniques. For discussions, I thank Margrit Fooke and Frank Hentrich.

In addition, I am grateful to Ernst Leitz, GmbH, Cologne, for lending to me a Leica SL camera, with which all pictures of devices and manipulations for this contribution were made.

V. MANUFACTURERS

Microtomes and cryomicrotomes

DuPont Instruments, Sorvall Operations, Peck's Lane, Newton CT 06470, U.S.A.
LKB-Produkter AB, S-161 25 Bromma, Sweden.
C. Reichert AG, Hernalser Hauptstrasse 219, A-1170 Wien, Austria.

Refrigerated centrifuges

DuPont Instruments, Sorvall Operations, Peck's Lane, Newton CT 06470, U.S.A.

Gyrotory water bath shaker

New Brunswick Scientific Co., Inc., 1130 Somerset St., New Brunswick, NJ 08903, U.S.A.

Low-temperature water bath

Gebrüder Haake, Goerzallee 249, D-1000 Berlin 37, FRG.

Bacteria counter

Coulter Electronics, 590 W 20 St., Hialeah, FL 33010, U.S.A.
Hausser Scientific, Township Line Rd., Blue Bell, PA 19422, U.S.A.

Stirrer, "Vortex"

Scientific Industries Inc., 70 Orville Dr., Bohemia, NY 17716, U.S.A.

Variable-speed stirrer for embedding mixtures

Talboys Engineering, 8 Palisade Av., Emerson, NJ 07630, U.S.A.

Grating replica

Ernest F. Fullam, Inc., P.O. Box 444, Schenectady, N.Y. 12301, U.S.A.

A general list of suppliers of material and equipment for transmission electron microscopy is published in "Practical Methods in Electron Microscopy" (A. M. Glauert, ed.), Vol. 3, pp. 339–347 (1974), see References.

REFERENCES

Aberman, R., Salpeter, M. M., and Bachmann, L. (1973). *In* "Principles and Techniques of Electron Microscopy", Vol. 2, pp. 197–217 (Ed. M. A. Hayat). Van Nostrand Reinhold Company, New York, Cincinnati, Toronto, London and Melbourne.

Agar, A. W., Alderson, R. H., and Chescoe, D. (1974). *In* "Practical Methods in Electron Microscopy", Vol. 2, pp. 1–337 (Ed. A. M. Glauert). North-Holland Publishing Company, Amsterdam and London. American Elsevier Publishing Co., Inc., New York.

Alderson, R. H. (1975). *In* "Practical Methods in Electron Microscopy", Vol. 4, pp. 1–124 (Ed. A. M. Glauert). North-Holland Publishing Company, Amsterdam and London. American Elsevier Publishing Co., Inc., New York.

Anonymous (1976). "The American Type Culture Collection". American Type Culture Collection, Rockville.

Anonymous (1974). "Deutsche Sammlung von Mikroorganismen. German Collection of Microorganisms. DSM." Gesellschaft für Strahlen- und Umweltforschung mbH, München.

Anonymous (1953). "Difco Manual of Dehydrated Culture Media and Reagents for Microbiological and Clinical Laboratory Procedures." Difco Laboratories, Inc., Detroit.

Anonymous (1975). "Instructions for Embedding Tissue in Epon 812 Based on Luft's Procedure." Ladd Research Industries, Inc., Burlington, Vt.

Anonymous (1968). "Mikrobiologische Untersuchungsmethoden." E. Merck AG, Darmstadt.

Bachmann, L., and Schmitt-Fumian, W. W. (1973). *In* "Freeze-Etching Techniques and Applications", pp. 73–79 (Eds. E. L. Benedetti and P. Favard). Société Française de Microscopie Électronique, Paris.

Bullivant, S. (1973). *In* "Advanced Techniques in Biological Electron Microscopy", pp. 67–112 (Ed. J. K. Koehler). Springer-Verlag, Berlin, Heidelberg and New York.

Dubochet, J. (1973). *In* "Principles and Techniques of Electron Microscopy", Vol. 3, pp. 115–151 (Ed. M. A. Hayat). Van Nostrand Reinhold Company, New York, Cincinnati, Toronto, London and Melbourne.

Dupouy, G. (1968). *In* "Advances in Optical and Electron Microscopy", Vol. 2, pp. 167–250 (Eds. R. Barer and V. E. Cosslett). Academic Press, London and New York.

Edie, J. W., and Karlsson, U. L. (1972). *J. Microscopie*, **13**, 13–30.

Fooke-Achterrath, M., Lickfeld, K. G., Reusch, V. M., Jr., Aebi, U., Tschöpe, U., and Menge, B. (1974). *J. Ultrastruct. Res.*, **49**, 270–285.

Glauert, A. M. (1974). *In* "Practical Methods in Electron Microscopy", Vol. 3, pp. 1–198 (Ed. A. M. Glauert). North-Holland Publishing Company, Amsterdam and London. American Elsevier Publishing Co., Inc., New York.

Hagen, P.-O. (1971). *In* "Inhibition and Destruction of the Microbial Cell", pp. 39–76 (Ed. W. B. Hugo). Academic Press, London and New York.

Hawkes, P. W. (1973). "Electron Optics and Electron Microscopy". Taylor and Francis Ltd., London.

Kellenberger, E., Schwab, W., and Ryter, A. (1956). *Experientia*, **12**, 421–422.

Kellenberger, E., Ryter, A., and Séchaud, J. (1958). *J. Biophys. Biochem. Cytol.*, **4**, 671–678.

Kellenberger, E., and Ryter, A. (1964). *In* "Modern Developments in Electron Microscopy", pp. 335–393 (Ed. B. M. Siegel). Academic Press, New York and London.

Kellenberger, E., Séchaud, J., and Blondel, B. (1972). *J. Ultrastruct. Res.*, **39**, 606–607.

Koehler, J. K. (1973). *In* "Principles and Techniques of Electron Microscopy", Vol. 2, pp. 53–98 (Ed. M. A. Hayat). Van Nostrand Reinhold Company, New York, Cincinnati, Toronto, London and Melbourne.

Lickfeld, K. G., Achterrath, M., and Hentrich, F. (1972). *J. Ultrastruct. Res.*, **38**, 279–287.

Lickfeld, K. G., Hentrich, F., and Fooke-Achterrath, M. (1975). *Arzneim.-Forsch.* (*Drug Res.*), **25**, 456.

Lickfeld, K. G., Almert, U., and Menge, B. (1976a). *Microscopica Acta*, **77**, 441–444.

Lickfeld, K. G., Menge, B., Hohn, B., and Hohn, T. (1976b). *J. Mol. Biol.*, **103**, 299–318.

Luft, J. H. (1961). *J. Cell Biol.*, **9**, 409–414.

Luft, J. H. (1973). *In* "Advanced Techniques in Biological Electron Microscopy", pp. 1–34 (Ed. J. K. Koehler). Springer-Verlag, Berlin, Heidelberg and New York.

Meek, G. A. (1971). "Practical Electron Microscopy for Biologists." Wiley-Interscience, London, New York, Sydney and Toronto.

Millonig, G., and Marinozzi, V. (1968). *In* "Advances in Optical and Electron Microscopy", Vol. 2, pp. 251–341 (Eds. R. Barer and V. E. Cosslett). Academic Press, London and New York.

Mühlethaler, K., Hauenstein, W., and Moor, H. (1973). *In* "Freeze-Etching Techniques and Applications", pp. 101–106 (Eds. E. L. Benedetti and P. Favard). Société Française de Microscopie Électronique, Paris.

Nanninga, N. (1973). *In* "Freeze-Etching Techniques and Applications", pp. 151–179 (Eds. E. L. Benedetti and P. Favard). Société Française de Microscopie Électronique, Paris.

Nash, T. (1966). *In* "Cryobiology", pp. 179–211 (Ed. H. T. Meryman). Academic Press, London and New York.

Oberzill, W. (1967). "Mikrobiologische Analytik". Verlag Hans Carl, Nürnberg.

Okinaka, H. T. (1971). Thesis. University of California, Los Angeles.

Parsons, D. F., Matricardi, V. R., Sujeck, J., Uydess, I., and Wray G. (1972). *Biochim. Biophys. Acta*, **290**, 110–124.

Plattner, H., Schmitt-Fumian, W. W., and Bachmann, L. (1973). *In* "Freeze-Etching Techniques and Applications", pp. 81–100 (Eds. E. L. Benedetti and P. Favard). Société Française de Microscopie Électronique, Paris.

Porter, K. R., and Fotino, M. (1974). *In* "Electron Microscopy 1974", Vol. II, pp. 166–167 (Eds. J. V. Sanders and D. J. Goodchild). The Australian Academy of Science, Canberra.

Rebhun, L. I. (1973). *In* "Principles and Techniques of Electron Microscopy", Vol. 2, pp. 3–49 (Ed. M. A. Hayat). Van Nostrand Reinhold Company, New York, Cincinnati, Toronto, London and Melbourne.

Reid, N. (1974). *In* "Practical Methods in Electron Microscopy", Vol. 3, pp. 217–347 (Ed. A. M. Glauert). North-Holland Publishing Company, Amsterdam and London. American Elsevier Publishing Co., Inc., New York.

Reimer, L. (1967). "Elektronenmikroskopische Untersuchungs- und Präparationsmethoden." Springer-Verlag, Berlin, Heidelberg and New York.

Reynolds, E. S. (1963). *J. Cell Biol.*, **17**, 208–212.

Ryter, A., Kellenberger, E., Birch-Andersen, A., and Maaløe, O. (1958). *Z. Naturforsch.*, **13b**, 597–605.

Séchaud, J., Ryter, A., and Kellenberger, E. (1959). *J. Biophys. Biochem. Cytol.*, **5**, 469–478.

Sjöstrand, F. S. (1967). "Electron Microscopy of Cells and Tissues". Academic Press, New York and London.

Vennrooij, G. E. P. M. (1975). *Arzneim.-Forsch.* (*Drug Res.*), **25**, 451.

Wischnitzer, S. (1973). *In* "Principles and Techniques of Electron Microscopy", Vol. 3, pp. 3–51 (Ed. M. A. Hayat). Van Nostrand Reinhold Company, New York, Cincinnati, Toronto, London and Melbourne.

CHAPTER V

Electron Microscopy of Small Particles, Macromolecular Structures and Nucleic Acids

D. KAY

Sir William Dunn School of Pathology, University of Oxford, Oxford

I. INTRODUCTION

The resolving power of current production transmission electron microscopes is about 0·2 nm as evidenced by measurements on lattice patterns obtained with inorganic crystalline specimens. On the other hand the resolution which can be demonstrated in biological specimens is about 2 nm or one order of magnitude poorer. For example, Horne *et al.* (1975) have demonstrated a resolution of 2 nm in images of large crystalline arrays of the small isometric plant virus, brome grass mosaic, by the use of optical diffraction methods. This discrepancy could mean that specimen preparation methods are lagging seriously behind the development of the electron

8

microscope or that there is some intrinsic but as yet imperfectly understood reason why biological material should not demonstrate detail of less than 2 nm. It is possible that improved specimen preparation methods together with advanced techniques for image processing will result in demonstrable resolution in biological specimens tending towards that which the microscope is clearly able to achieve but for the time being the vast majority of high resolution electron microscopy of biological material will be carried out by the negative contrast procedure in its various forms.

The negative staining or negative contrast method has been successfully applied to an enormous variety of small macromolecular structures such as viruses of all types, bacterial cells and their appendages, flagellae and pili, membranes of many varieties of cells and protein molecules both individually and in ordered arrays either naturally occurring, as in collagen, or as laboratory produced crystals.

In order to obtain a useful image of a negatively stained preparation the specimen has to contain at least a proportion of protein or lipid. Nucleic acids on the other hand, when stripped of protein, do not give good images with negative staining methods. The central importance of nucleic acids to biology has resulted in much effort being spent on preparing specimens which would yield meaningful electron microscope images. The most useful method has been that devised by Kleinschmidt and Zahn (1959) which, in one or other of its developments, makes it possible to examine all forms of nucleic acid although not at high resolution. Although the double helix of DNA is only about 2 nm in diameter the Kleinschmidt method does not require the use of a high resolution electron microscope because it increases the cross-sectional diameter of the strands by up to a factor of 10 thus rendering them visible at quite modest magnifications and resolving powers easily obtained by most microscopes in use today. The purposes to which this method have been put are very numerous and are constantly increasing.

In addition to the "descriptive" studies which have been made by the negative staining and Kleinschmidt techniques there has been a steady development of methods whereby the location of specific chemical and functional sites on macro-molecular structures can be visualized. For example, the sites of attachment of repressor molecules to DNA strands, the sites of action of specific restriction endonucleases and "unwinding" proteins and the location of particular antigenic complexes on viruses and cells can now be studied with considerable precision.

The function of this Chapter is to describe the most successful methods now in use for negative staining and for Kleinschmidt-type specimen preparation together with the variations which have proved useful in special instances. Details will also be given for the preparation of specimen grids adapted to these methods. It is not the aim to give a complete review of electron microscope methods applicable to the study of small molecules and macromolecular structures and it is quite possible that some workers' favourite methods have been omitted and others thought by some to be of slight value have been included. This is inevitable in a Chapter which includes many methods used by the author or by his colleagues and other methods devised by experimenters whose work has enormously advanced the study of the fine structure of biological material. No attempt has been made to rigorously attribute the methods to their originators and indeed these may not be known in all cases. The references given are a general guide into the literature dealing with particular methods where the results clearly indicate the effectiveness of the procedure.

A. Preparation of specimen grids

For all high resolution work it is necessary to use carbon films to support the specimens. The alternative films made of organic polymers such as cellulose nitrate (e.g. collodion and Parlodion, which is a trade name for a variety of collodion) or formvar (polyvinyl formaldehyde) are satisfactory for low magnification use but are insufficiently stable at high magnification and their use will vitiate all attempts at high resolution electron microscopy. Carbon films can be used to stabilize collodion or formvar films and the double film referred to as carbon-coated formvar or collodion is commonly used for medium to high magnification work. For the very highest resolution, which demands high magnification, the carbon-coated plastic film is generally too thick and is usually replaced by plain carbon films of extreme thinness.

There are some circumstances where plastic films have been noted to be preferable to carbon films. For example Delius et al. (1972) have observed that the air dried surface of Parlodion films is superior to the side formed in contact with the water surface in respect of background coarseness and better contrast for preparing DNA by the Kleinschmidt method. However the general opinion is that carbon-coated plastic films are perfectly satisfactory for this purpose.

1. Carbon-coated plastic films

The general procedure is to prepare grids coated with either collodion or formvar and then evaporate onto the plastic film a thin layer of carbon. Collodion films are made by allowing a drop of a 2–3% solution of collodion (or Parlodion) in amyl acetate to fall from about 1 cm onto the surface of distilled water in a dish of 10–15 cm diameter. The drop spreads over the surface and, after the solvent has evaporated, a thin film of plastic remains floating on the surface. The water should be double distilled and care should be taken to eliminate dust by working in as dust-free an area as possible and by cleaning the water surface by casting and removing several plastic films. Grids are then applied to the floating film with their convex sides downwards. A piece of thin tissue paper is then applied to the floating film carrying the grids. The film adheres to the paper which can be lifted bearing the grids sandwiched between it and the film. After drying under an electric lamp the grids are stored, still attached to the paper, until required.

Formvar films are cast on glass microscope slides by dipping them into a 0.3% solution of formvar in 1 : 2 dichloroethane. The slides are allowed to dry, standing vertically under a cover. The films are scored with a needle about 1 mm from the edge and are floated off by lowering the slides at an angle into a bath of distilled water. It is essential that the slides are not overcleaned otherwise the films will not strip off. A commonly used procedure is to clean the slides in soap and water, dry, and grease the surface by rubbing the fingers over it. After polishing with a cloth kept specially for the purpose the slides are coated in formvar and stripped as described above. Grids are applied to the floating film as described for collodion films. Wells (1974) has suggested polishing the slides with the household metal polish, "Duraglit", or fine alumina powder.

Some grids from each batch should be examined in the electron microscope for cleanliness, thinness and freedom from holes. There is generally no need to make a measurement of film thickness but a guide to satisfactory thickness can be obtained on microscopes equipped with an electronic exposure measuring device by setting the brightness of the screen to full scale deflection and then introducing a filmed grid into the beam. The screen brightness should not diminish to less than 95% at 60 kV. If the films are too thick the concentration of plastic in the solvent should be reduced and *vice versa*.

The grids are coated with carbon by placing the paper sheet carrying them in a vacuum evaporating apparatus (see p. 185) at a distance of 12–15 cm from the arc carbons, evacuating the workchamber to 10^{-4} Torr and evaporating as much carbon as required. An indication of the amount of carbon evaporated can be obtained by placing a sheet of white paper at the same level as the grids and observing the discoloration of the paper as the evaporation proceeds. Usually a faint browning is all that is needed. If the test paper is partially covered with an object, such as a small coin, the colour of the carbon layer can be clearly seen when the paper is removed and it can be kept as a record for that batch of grids. Deposition of carbon can be followed more precisely by the use of commercially made thickness monitors or by a device of the type described by Griffith (1973) which consists of a 22 mm × 22 mm glass cover slip held between two insulated supports and mounted close to the grids being coated. Leads are connected to opposite sides of the cover slip with silver conductive paint and are taken to a resistance meter. Before evaporation the resistance is about 5×10^7 ohms and this falls to 5×10^4 ohms when a suitably thin film has been deposited.

2. Plain carbon films

These can be made by two general procedures. The carbon-coated plastic films mounted on grids can be treated with the same solvent in which the underlying plastic was dissolved so as to remove the plastic and leave the carbon film. Grids can be laid on a stainless steel mesh and lowered into the solvent for an hour or so with little or no agitation as the carbon films are very easily broken by movement through liquid or through air. Another method to remove the plastic film is to place the grids on a small stack of filter paper immersed in solvent which is changed three times over a period of 3 h.

An alternative method of preparing carbon films is to evaporate the carbon onto the very clean surface of freshly cleaved mica. Griffith (1973) suggests that the coated mica be kept in a humid atmosphere for 1 h and then stripped off by dipping into water. Frequently it is found that the carbon layer is difficult to strip. Johansen (1974) recommends that the cleaved mica be kept for two days in a dust-free cabinet containing a vessel of water before evaporation of carbon which can then be easily stripped off.

Carbon films prepared in this way are very fragile and usually do not adhere well to the grids. Adhesion can be improved by treating the grids with a 1% solution of polybutene 6000 in xylene or as suggested by Bradley (1965) with a solution made by dissolving the adhesive from Sellotape in chloroform. Sometimes etching the grids for a few seconds in concentrated hydrochloric acid improves adhesion of carbon films. Johansen (1974) has described a method of making grids with a hydrophilic surface to which carbon films adhere well. This procedure is described in the section on perforated films (p. 184). The floating carbon films are applied to grids by placing the grids, on a piece of stainless mesh, under the film and gently lowering the water level.

It is possible to make carbon films down to 2 nm thick by the method described above but the all-important surface properties of the films depend to a large extent on the properties of the vacuum evaporation equipment in use. Carbon films tend to be hydrophobic and in consequence do not permit aqueous specimens to spread on their surfaces and frequently do not allow particles to adhere to them. These faults can be overcome by special treatments (see below) but in addition carbon films have an unexpectedly rough surface. This may be due to the vacuum equipment contributing a layer of oil from the vapour backstreaming from the diffusion pump.

This can be minimized by using a well-baffled pump fitted with a liquid nitrogen-cooled trap above the diffusion pump. Also the grids can be protected until just before carbon evaporation by a cover which can be removed by some form of external control. The problem of grid contamination in the vacuum evaporator has been studied by Horne and Pasquali Ronchetti (1974) who used a vacuum system consisting of a mechanical backing pump fitted with a liquid nitrogen trap backing a mercury diffusion pump which itself was followed by a nitrogen-cooled trap. The vacuum during carbon evaporation was 2×10^{-6} Torr.

The problem of diffusion pump fluids contaminating the surfaces of evaporated films can be overcome by the use of a pumping system which uses ion pumps instead of diffusion pumps. Johansen (1974) has described a system for making ultrathin carbon films using an ion pumped vacuum apparatus in which a controlled amount of carbon is evaporated from an arc not directly onto the substrate, in this case cleaved mica, but indirectly by reflection from the sides of a glass cylinder. The mica is placed 120 mm below the carbon rods and is surrounded concentrically by a glass cylinder 125 mm diameter and 80 mm high. It is protected from direct carbon evaporation by a circular stop 20 mm diameter supported by wires at the level of the top of the cylinder and concentric with it and the arc. The carbon rods are machined down to 1 mm diameter at their ends and are pressed together by a spring. A mechanical stop prevents the carbon rods from moving more than a set distance and so limits the amount of carbon evaporated. The method is capable of giving carbon films down to 1 nm thick with very low surface structure. A vacuum of 10^{-5} Torr is used.

Ultrathin films cannot be supported directly on standard grids of even the finest mesh and are always deposited on perforated supporting films (holey films) which themselves are placed on the standard grids.

Fernandez-Moran et al. (1966) have also described an oil-free pumping system consisting of a Varian Vac-ion pump, liquid nitrogen and liquid helium traps and Vac-sorption fore pumps which gave a vacuum of 10^{-8}–10^{-9} Torr.

3. Perforated support films

Grids bearing films containing small holes (holey films) have many uses in electron microscopy. Carbon-coated holey films are used to detect and correct astigmatism in the objective lens. Provided that the holes are small enough for a complete high magnification image to be seen through the viewing binoculars an underfocused Fresnel fringe can be observed and by adjustment of the astigmatism correction device and the fine focusing a uniformly thin fringe can be observed which should vanish around the whole circumference as true focus is reached. Complete correction of astigmatism is of course never achieved but the aim of the operator is to make the adjustment of the astigmatism controls so that there is as little difference as possible between the focusing currents at which the fringe begins to disappear across one diameter and finally disappears across another diameter at right angles to the first.

Another use for holey films is to support droplets of negatively stained specimen within the holes. The specimen is then held around the edges of the hole and can be viewed without any support film being interposed.

A third use for holey films is to support ultrathin carbon films which are too weak to bridge the grid squares of standard support grids. Many methods have been published for making perforated films, some of which exert a degree of control over the size of the holes. The proportion of the film occupied by holes, of whatever size, can be varied over wide limits so that films can be produced that have only a few holes or that have the appearance of a net.

A simple but uncontrolled method of making perforated films is to dip a slide into a solution of formvar as used for making standard formvar films and then to breathe onto the slide before the solvent has evaporated. The condensed moisture in the form of fine droplets causes holes to appear in the film which can be stripped and mounted on grids in the usual way. A more controlled way has been described by Harris (1962). Glycerol at 1 : 8 to 1 : 120 is added to a 0·25% formvar solution in 1 : 2 dichloroethane and shaken vigorously before being used to cast films on glass slides. After drying the slides are held in a jet of steam for a few moments before being stripped in the usual way. The fine glycerol droplets are washed out by the steam leaving the required holes, the size of which depends on the proportion of glycerol.

Fukami and Adachi (1965) have prepared perforated films, which they call microplastic nets or microgrids, by first treating cleaned glass slides with an aqueous solution of the cationic surface active agent di-stearyl dimethyl ammonium chloride (0·03%) followed by washing in water and drying in air. The slides are cooled in a refrigerator for 3–50 sec to below the dew point of the room atmosphere and flooded with a solution of Triafol (cellulose acetobutyrate) 0·1–0·5% in ethyl acetate. The slides are inclined and allowed to dry. It is important that the relative humidity be kept within the range of 50–70%. If it is below 50% the Triafol solution can be cooled or if above, the work can be carried out under an infrared lamp. The slides are then treated with a solution if a hydrophilic agent, sodium dialkyl sulphosuccinate (0·5%) in water for 3–10 min, washed in water and then stripped by lowering at an angle of 10° into a bath of water. If the films do not strip a prolonged treatment with the hydrophilic agent is suggested.

The films can be made with holes from 0·1–0·6 μm across in which the majority of the film area is perforated.

The holey plastic films made as described above can be used to support thin carbon films made by the method of evaporation onto mica, but for the highest stability, Johansen (1974) recommends that the plastic net be used only as a temporary support for a net made of superimposed layers of carbon, gold and silicon monoxide because the plastic itself is not sufficiently rigid when subjected to high brightness electron beams. Micro-plastic grids made by the procedure of Fukami and Adachi (1965) and supported on 200 mesh copper grids were coated with a layer of evaporated carbon 15–20 nm thick which was then covered with a layer of evaporated gold of similar thickness. A layer of silicon monoxide evaporated from a basket placed 100 mm from the grids was then formed. The purpose of this layer was to give a hydrophilic surface to which the thin carbon films would adhere. The Triafol was then removed by treating the grids with ethyl acetate on a pad of filter paper. The grids were used to pick up ultrathin films of carbon which had been prepared by indirect evaporation onto mica (see p. 182).

B. Surface treatment of carbon films

Carbon films are usually hydrophobic, sometimes extremely so, with the result that specimens fail to spread on their surfaces and particles fail to adhere. There are several methods by which this can be overcome. The simplest is to irradiate the grids with ultraviolet light from a germicidal lamp placed at a distance of 20–30 cm for 30 min or more. The change in surface properties is not permanent and it is suggested that carbon-filmed grids be irradiated just before use if difficulties are encountered with non-wettability. The mounting of strands of DNA in the absence of protein has been made possible by activating the carbon film surface or giving it a charged surface. A suitable method (Griffith, 1973) is to subject the carbon-filmed grids to a high voltage discharge in the presence of oil vapour. The pressure in the workchamber of a vacuum evaporation apparatus is allowed to rise to 100–300 μmHg while being pumped by a diffusion pump loaded with Kinney Super X oil. The grids are placed between two insulated horizontal metal plates about 5 cm apart which are connected to a source of 10–20,000 V a.c. The grids retain their ability to bind DNA strands for only 20 min.

Another method used to prepare charged carbon films suitable for adhesion of DNA strands in the absence of protein (e.g. cytochrome c, as in the Kleinschmidt method) has been described by Dubochet (1971). The grids were subjected to a glow discharge of 500 V a.c. (130 V/cm) in a vacuum at 0·1–0·15 Torr in the presence of amylamine.

II. METAL SHADOWING

The oblique evaporation of a thin layer of heavy metal onto filmed grids carrying all manner of specimens was the earliest method for generating sufficient contrast to give useful images in the electron microscope. The limitation of the method is that the metal layer has a structure of its own which is clearly visible at high magnification and obscures the finer details of the specimen. This structure is due to the formation of microcrystals of the metal. Possibly the smoothest and least crystalline evaporated films are those prepared by the simultaneous evaporation of platinum and carbon (Bradley, 1959). The problem of evaporated metal crystallinity was side-stepped by the development of negative contrast methods (see p. 187) in which the structure of the staining materials is sufficiently fine as to be negligible at all but the highest magnifications.

Metal shadowing has retained a very useful role at lower magnifications where it is used to give contrast to strands of nucleic acid prepared by the Kleinschmidt method and to specimens prepared by freeze fracturing and freeze-etching (methods discussed in Chapter IV). The metal may be evaporated obliquely from one direction, or from two directions at right angles or from all directions by rotary shadowing. The most frequently used metals are platinum, gold and palladium, all noble metals, singly or in pairwise combination. The assumption is that the metal atoms emanating in straight lines from the high temperature source, which can be considered as a point, strike the grid surface or the specimen and remain where they hit apart from slight aggregational movements to form the microcrystalline structure which is always present. This assumption is probably true in the case of the unreactive noble metals but is not true when a reactive heavy element is evaporated. Griffith (1973) has described a method for shadowing with tungsten in which the evaporated metal is selectively nucleated by different parts of the specimen to produce a new form of contrast which is described as vapour phase positive staining.

A. Oblique evaporation of noble metals

It is assumed that a vacuum apparatus is available which has a work-chamber, preferably of glass for ease of observation, ideally about 30 cm in diameter and about the same in height. The chamber should be equipped with lead-ins to carry heavy currents of up to 60 A and others suitable for light currents at high voltages of up to 10 kV. For rotary shadowing it is useful to have a mechanically driven shaft entering the chamber and rotating in a horizontal plane. The pumping system should be capable of reaching 2×10^{-6} Torr and should have a liquid nitrogen-cooled trap between the diffusion pump and the chamber. In order to reduce vibration the mechanical pump should be mounted on the floor, not on the frame of the apparatus.

Grids bearing specimens to be shadowed are placed on the work table of the chamber at a distance of 12–15 cm from the source. For most work with nucleic acid strands the angle of shadowing used is about 7° to the plane of the grid. Clearly there is the possibility that the film surface is not flat or that the grid itself is bent in which case it is impossible to be certain of the shadowing angle. Many devices have been designed to hold grids flat. They consist of metal plates with recessed holes onto which the grids are placed while they are held flat by superimposed metal plates with corresponding holes. The difficulty with these devices is that they often prevent large areas of the grids from being shadowed when the angle of incidence is as low as 7°. A method of overcoming grid unevenness is to use nickel grids instead of the usual copper grids and to hold them flat by the use of a magnet underneath a thin glass work surface. The magnetic properties of the nickel grids apparently do not cause any problems when they are placed in the electron microscope.

The metal to be evaporated, in the form of wire 0·1 or 0·2 mm diameter, is coiled round the midpoint of a piece of tungsten wire 0·5 or 1·0 mm diameter bent into a "V" shape of about 5 mm sides with the tops of the arms extended sideways about 2 cm. The tungsten wire filament is clamped between sturdy metal arms bolted to the heavy current lead-ins of the apparatus. Gold, platinum and palladium can be obtained in wire form as alloys of various compositions, e.g. 40% palladium and 60% gold, 20% palladium, and 80% platinum. Gold–palladium alloys are easy to evaporate as they have a low melting point. Pure platinum is more difficult because of its high melting point. All these metals form alloys with the tungsten wire filament eventually weakening it and causing it to crack or burn out. Tungsten, once it has been heated to a high temperature, becomes very brittle and is easily broken by a touch or by thermal contraction. It is therefore advisable to use a fresh tungsten filament for each evaporation.

Evaporation is carried out by pumping the chamber to the required vacuum and gently heating the filament by steadily raising the current passed through it until the metal melts to form a droplet and slowly evaporates. The tungsten filament has to be heated to white heat but this can be comfortably viewed by eye through a filter made by fogging a piece of film, developing and fixing. The vacuum required depends on the fineness of the metal shadow aimed at but should not be less than 2×10^{-5} Torr.

B. Rotary evaporation

When nucleic acid strands are shadowed from one direction only the visibility of the strands depends on the angle at which they lie towards the source of the metal. Where they point directly to the source they are almost invisible. This disadvantage can be partially overcome by repeating the shadowing at right angles to the first direction. A better method is to rotary shadow the specimens. The grids should be arranged in a circle 1–2 cm diameter and held in a holder which clamps them flat. The holder should be arranged so that it can be rotated at 60–120 rpm with its centre of rotation at such a position that the filament is 10–12 cm away and the angle of shadowing is about 7° to the surface of the grids. Many devices have been made to rotate specimens. They have used torsion springs, clockwork motors and electric motors mounted inside the workchamber but the most satisfactory method uses an external motor driving through a rotary seal into the chamber.

C. Shadowing by selective nucleation

Griffith (1973) has described a method of shadowing with tungsten in which the metal is deposited preferentially on parts of the specimen which differ in chemical composition. A straight tungsten wire (0·5 mm diameter) is clamped between holders 3 cm apart in the vacuum chamber at a height of 8 cm above the specimen. The wire should be heated by passing such a current that it thins and breaks in 5–7 min. The specimen should be rotated but the angle is not critical and can be between 8° and 45°. The tungsten, which is chemically reactive, is deposited preferentially on protein, less on DNA and least on the carbon support film.

III. NEGATIVE CONTRAST METHODS

Negative contrast or negative staining is so called because the specimen is seen as a light object against a dark background. In electron microscope terms the biological specimen, being composed mainly of elements of low atomic number, is electron transparent and is outlined by the negative staining material, which is always composed of salts of elements of relatively high atomic number and which appears electron opaque. The stain does not react chemically, at least to any great extent with the specimen but flows, into its surface relief and may penetrate inside it if there are openings in its structure. The negative staining materials in general use are sodium or potassium phosphotungstate and uranyl acetate in aqueous solutions, usually at defined pH's. Many other substances have been used for negative staining, ammonium molybdate, uranyl formate, uranyl magnesium acetate, potassium silicotungstate, sodium tungstate, but are comparatively little used.

Negative staining is not a precise procedure as there are many variables which cannot easily be controlled. Carbon-coated plastic films are most commonly used to support the specimens or, for the highest resolution, very thin carbon films. The surfaces of these films are not uniformly wettable. Specimens have to be suspended in a fluid, which can be distilled water, or solutions of buffers or of volatile salts such as ammonium acetate. Specimens may be fractions taken from density gradients and may contain high concentrations of sucrose or caesium chloride. There may be detergent substances present which have been used for breaking open cells or cell organelles. All these substances together with the nature of the specimen itself affect the spreading of the stain on the grid and its consequent drying pattern and thickness. As a result many grids may be of low quality because the stain is too thick or too thin or the specimen may have adhered in too large or too small an amount. It is therefore worthwhile making several dilutions of each specimen and trying different staining methods and materials. It is also necessary to examine several grid squares across the width of the grid in order to locate well stained areas.

Negative staining has been applied to a wide variety of specimens. Large specimens such as bacterial cells, although too thick to show internal detail, frequently show considerable surface detail. Bacterial appendages, flagella and pili, are very well delineated whether attached to the cells or free. Viruses of all kinds, animal, plant, insect and bacterial have been examined by negative staining and have yielded images of great detail. Cellular components such as ribosomes, polysomes, microtubules and membranes have also been successfully examined. Protein molecules both singly and associated in functional groupings have been examined by this method which is generally applicable to small particulate specimens over a wide range of size.

A. Negative staining materials

1. Phosphotungstic acid

This is used at 1 or 2% in aqueous solution and is adjusted with sodium or potassium hydroxide to pH's between 6 and 7. Lower pH's down to 4 have been tried.

2. Uranyl acetate

This is used at 1 or 2% in aqueous solution but is not usually adjusted in pH which is about 4·0. The pH may be raised to 6·0 by the addition of concentrated ammonia without causing precipitation.

3. Uranyl formate

A 1% solution in water has a pH of 3·7. It has been claimed to give less granularity than the acetate.

4. *Ammonium molybdate*

This has been used at concentrations from 3·5% to 10% with its pH adjusted to be in the range 5·2–7·5.

Stains should be made up in double distilled water in glass vessels which have been well cleaned. Some staining materials dissolve easily but others may need prolonged stirring. This can conveniently be done on a magnetic stirrer with a small magnetic "flea" coated in polythene. Substances which are difficult to dissolve may be gently warmed in a water bath at a temperature not exceeding 60°C. Undissolved material should be removed by filtration through hard filter paper. It is rarely necessary to use finer filters, e.g. Millipore, but if trouble is experienced with "foreign" particles on the grids very fine filters could be tried. There is no need to store staining solutions in the cold but uranyl solutions should be protected from light.

B. Methods

1. *Spray methods*

One of the earliest methods of preparing negatively stained specimens was to mix equal volumes of the suspension of the specimen with 2% potassium phosphotungstate and spray the mixture onto carbon-filmed grids. The apparatus used for spraying was a commercially available nasal spray which was fitted with glass bulb and a tube bent downwards at a right angle to the spray outlet so as to catch the larger droplets and allow the smaller to pass and fall onto the grids. It was necessary that the specimen be made up in water or a volatile buffer. Ammonium acetate, ammonium bicarbonate and collidine acetate at various concentrations from 0·01–0·2M have been used. If the specimen contained much non-volatile material apart from the stain this would crystallize on the grid and spoil the preparation. The spray method was therefore somewhat inconvenient in that it required a special spray gun and the specimen had to be freed from non-volatile material either by dialysis or centrifugation and washing and it produced an aerosol which might be unacceptable if the specimen were pathogenic. Special devices have been made to overcome this latter objection (Horne and Nagington, 1959). The method does possess one very valuable asset. All the particles of the specimen irrespective of size or ability to adhere to the grid are deposited equally on its surface. This means that relative counts of different particles in the specimen can be made and an absolute particle count can be made by incorporating into the sprayed mixture a known quantity of polystyrene latex particles.

2. *Droplet-on-grid methods*

A very simple method which will deal with a wide variety of specimens, including those with large amounts of non-volatile substances, performs all operations on the grid surface. A suitable coated grid is held by its edge in a pair of forceps the tips of which are held closed by pushing a small elastic band along the forcep legs. A droplet of the specimen in the fluid in which it was prepared, or suitably diluted in water or buffer, is placed on the grid and left for a period ranging from a few seconds to a minute. The bulk of the fluid is then removed by touching the grid to the edge of a torn piece of filter paper. It is important that not all the fluid is removed and the grid should not be allowed to dry. The grid is now held over a waste receiver such as a small beaker and two or three droplets of stain are allowed to run over it in quick succession. The droplets replace the fluid in which the specimen was applied and surround it with stain. If carbon-coated plastic grids made and treated as described on p. 184 are used it will be found that adequate amounts of the specimen will adhere to the grid. The excess stain is removed with filter paper.

It is often found that particles are unevenly distributed over the surface of the grid so that some areas are heavily covered while others are left bare. This is unsatisfactory when the proportion of particles by size, appearance or some other criterion is being studied. A method for improving the distribution of particles has been described by Kirkpatrick et al. (1969) for microtubules derived from brain. The microtubules suspended in a buffer consisting of hexylene glycol (131 ml) and KH_2PO_4 (454 mg) adjusted to pH 6·4 with KOH were applied as a droplet to a carbon-filmed grid. Most of the fluid was drained off with filter paper and a droplet of bovine serum albumin (1 mg/ml) was applied. This was again drained off and followed by a droplet of 2% phosphotungstate at pH 6·2 which was again drained off against filter paper.

3. Inverted droplet method

Nermut (1972) has examined a variety of animal viruses by placing carbon-coated grids on droplets of virus suspension lying on the surface of a sheet of dental wax. The grids are kept upside down while the specimens are washed and stained by transferring distilled water and then negative stain under them. They are finally dried by touching to filter paper. A somewhat similar method is to set out a row of droplets of specimen, washing fluid (water or buffer) and stain on a sheet of dental wax and place carbon-coated grids successively onto the drops. Droplet methods such as these avoid the necessity of holding the grids in forceps.

4. Immersion method

Bradley (1974) has examined piliated strains of *Pseudomonas aeruginosa* with adsorbed phage particles by immersing carbon-coated grids, held in forceps, in the cultures of the organism. After 2 min the grids were washed in water and then negatively stained by immersion in 0·67% neutral sodium phosphotungstate.

5. Pseudoreplica method

Horne et al. (1975) and Nermut (1975) have applied droplets of virus to freshly cleaved mica and, after allowing time for the particles to attach, the mica was washed and dried from the frozen state or air-dried. A thin layer of carbon was then evaporated onto the specimen. This was stripped off onto the negative staining solution which was either 4% sodium silico-tungstate (pH 6·5), 1% uranyl acetate or 3% ammonium molybdate (pH 6·5) and then picked up on grids. This type of specimen is termed a pseudo-replica because although the carbon film is a replica of the specimen the virus particles remain attached to the film and contribute to the image in the electron microscope. As an alternative to negative staining, the freeze-dried specimen on the mica may be shadowed with platinum–carbon at an angle of 35°–45° then with carbon at 90° and stripped off onto water.

6. Mica surface method with double negative stain

Horne and Pasquali Ronchetti (1974) have examined brome grass mosaic virus, potato X virus and adenovirus Type 5 at high concentration (7–18 mg/ml for brome grass mosaic virus) by mixing with either 3% am-monium molybdate (pH 5·2) or 2% potassium phosphotungstate (pH 6·7) and allowing it to dry in air on a sheet of freshly cleaved mica of rectangular shape 45 × 15 mm with one end cut to a point. A thin layer of carbon was then deposited on the mica at 2×10^{-6} Torr in an apparatus with well-trapped pumps (see p. 182). The carbon film was then floated off the mica onto a bath of a second negative stain which was either 2% uranyl acetate at pH 4·0–4·5 or 3% ammonium molybdate. This method was used to obtain very high resolution micrographs of large crystalline arrays of the virus particles which, when submitted to optical diffraction analysis could be shown to have a resolution of 2·0 nm. After floating on the second stain carbon-coated perforated plastic films were raised underneath the carbon film bearing the virus particles.

7. Surface spreading method

A droplet of a solution of the enzyme pyridine nucleotide transhydro-genase from Pseudomonas aeruginosa (Louie et al., 1972) was placed with a fine pipette onto the surface of a solution consisting of 0·1M potassium phosphate buffer (pH 7·5) containing 10mM 2-mercaptoethanol and 1mM EDTA. The solution had been dusted with talc particles to act as an indicator of the spreading of the protein monolayer. A grid prepared by the method of Fernandez-Moran *et al.* (1966, see p. 182) was placed on the surface for 1–2 min to allow the protein to diffuse to the grid surface. The grid was removed and the attached fluid absorbed with filter paper and then stained by floating on 1% uranyl formate or phosphotungstate at pH 6·6 for 2–5 min. After removal of excess stain the grids were air dried. These preparations yielded high resolution micrographs showing subunit structure in the basic 12×15 nm diameter particles of the enzyme.

8. *Methods using specimen fixation*

Certain types of specimen are destroyed or seriously altered by the negative staining procedure as described in the preceding sections. For example viruses which have lipid membranes such as the murine leukaemia virus show tail-like structures if they are not fixed before negative staining. Ribosomes and polysomes should also be fixed and attention paid to the ionic environment while fixation and negative staining are taking place. In a method of fixation described by Nermut (1972) droplets of virus suspension were placed on dental wax sheet and carbon-filmed grids laid upon them. After allowing the virus particles to attach to the grids they were transferred to droplets of fixative for 5–10 min, then to water and finally to negative stain. The fixative was either osmium tetroxide at 0·1–1·0% or 3–5% glutaraldehyde in phosphate buffered saline at pH 7·2. It is important to examine virus particles both with and without fixation and after treatment with different fixatives because the action of the fixatives themselves can alter the structure of the virus particles as in the case of the surface spikes of the influenza virus (Nermut, 1972).

Polysomes (polyribosomes) are affected by the ionic composition of the medium in which they are suspended. Magnesium ions in particular are essential for their integrity. Polysomes in fractions from sucrose density gradients can be prepared for electron microscopy by placing droplets in watch glasses standing in ice and floating carbon-coated plastic grids on them to allow the polysomes to attach. The grids are then transferred to droplets of fixative consisting of 0·25M formaldehyde (neutralized) and $1·5 \times 10^{-3}$M $MgCl_2$ for 5–10 min, also on ice. For negative staining the grids are inverted on droplets of uranyl formate or acetate at pH 4 also at 4°C.

Fernandez-Moran *et al.* (1966) have used a somewhat similar method to prepare specimens of the large protein molecules, the apohaemocyanins. A droplet of protein at 25–100 μg/ml in 0·01M ammonium acetate at pH 6·0 or 0·1M ammonium acetate with 0·0005M to 0·01M calcium or magnesium acetate at pH 6–7 is placed on perforated carbon-coated grids prepared by the procedure described on p. 184. After 1–2 min the grids are floated upside down on the same buffer for 15 sec and then transferred to a fixative containing 1–3% formaldehyde adjusted to pH 7·0 with KOH or 1·5%

glutaraldehyde in 0·01M phosphate buffer at pH 7·0 which is kept in an ice bath. After 15 sec the grids are rinsed in ammonium acetate buffer and then floated on a negative stain consisting of 0·5–2·0% phosphotungstate adjusted to pH 7·2 with KOH.

The lipid-coated bacteriophage PM2 has been examined by Hinnen et al. (1974) by spreading the phage suspension onto carbon-coated grids and touching them to the surface of a fixative consisting of 2·5% glutaraldehyde in 0·1M NaCl at pH 7·5. The grids were then negatively stained by touching to the surface of 2% uranyl acetate for 30 sec.

9. Effect of the pH of the stain

The effect of the pH of the stain on the appearance of the negatively stained specimen has been examined by Bradley (1961) and Nermut (1972). Bradley pointed out that some stains alter in pH during the course of drying. The enveloped virus of influenza shows marked differences if stained at pH 5·0 with phosphotungstate, when the surface spikes are well outlined, and at pH 7·5, when the stain penetrates the envelope and stains the nucleo-protein helix within.

Mellema et al. (1968) have examined the effect of pH on the negatively stained small globular protein molecules of chymotrypsin, papain, pepsin and trypsin and have compared the sizes of the particles seen in the electron microscope with the sizes determined by X-ray diffraction. The minimum sizes agreed with the X-ray data but the maximum sizes were 1·5–2·0 nm too high. It was concluded that only the average size and general shape of the molecules could be found by electron microscopy. The sizes were distorted most at pH's 2·0 and 6·0 with uranyl acetate and least at pH 4·2.

C. Positive staining

A specimen is considered to be positively stained when it, or regions of it, appear electron dense against an electron transparent background. It is the exact opposite of the negatively stained specimen. The method is little used because it rarely provides the high resolution given by the negative proced-ure. The staining materials are the same as are used for negative staining but negative contrast is avoided by washing away the excess stain before the specimen is allowed to dry.

Quite often a negatively stained specimen will be found to have areas which have become positively stained for example a bacteriophage where the heads appear as dark polygons and the tails are scarcely visible. The mechanism for this spontaneous and unwanted positive staining is not known but could be related to local pH changes as the stain dries or stresses introduced in the virus particles due to surface tension which could partially disrupt the virus capsid and allow entry of the stain to the nucleic acid core.

In order to stain a specimen positively it is applied to a carbon-filmed grid and time allowed for attachment as in negative staining. The stain, which may be 1% uranyl acetate or 1–2% phosphotungstate at neutral pH or lower is then applied and after a short period is washed off with water and the specimen is allowed to dry. A particularly informative example of the use of positive staining is given by Doyle *et al.* (1975) who by examining the pattern of darkly stained bands in various forms of collagen and making models based on the total amino-acid sequence of the tropocollagen monomer were able to arrive at the structure of collagen fibres. Another interesting example of positive staining is the depiction of a fine strand, probably of RNA, connecting the individual ribosomes in a polysome (Rich *et al.*, 1963). When polysomes are negatively stained the ribosomes are seen more clearly but the RNA strand is too thin to develop any negative contrast and is not seen.

IV. METHODS USING ANTIGEN–ANTIBODY COMPLEXES

Antibody molecules can be visualized in the electron microscope by negative staining. If, therefore, antibodies can be prepared against macro-molecular structures or their components then the location of the antigens, provided they are on the surface, can be identified. Yanagida and Ahmad-Zadeh (1970) prepared antiserum against purified coliphage T4D by repeated injections of the particles into rabbits. The serum was adsorbed with an extract of the uninfected host bacteria to remove antibacterial anti-bodies which are inevitably present no matter how well the phage has been purified. The serum was then adsorbed with what are termed defective lysates. These are the products of non-permissive host bacteria which have been infected with amber conditional lethal mutants of the wild-type phage. All the components of the phage are present in these lysates except the one coded for by the mutant gene. Adsorption of the serum containing anti-bodies to all the phage components with the defective lysate will remove all the antibodies except the one to the phage component coded for by the mutant gene. The adsorbed serum is therefore specific for that phage protein component. Several antisera were prepared and adsorbed with different defective lysates. These were tested against whole phage or phage-related particles. One drop of a phage suspension (5×10^{11} particles/ml) was placed on a carbon-coated grid and after 30 sec washed off with water. The grid was then floated, specimen side down, on a drop of the specific antiserum diluted 1 : 100 in phosphate buffer (Na_2HPO_4, 7 g; KH_2PO_4, 3 g; NaCl, 4 g; per litre of water containing $10^{-3}M$ NaN_3) in a small closed tube for 30 min to 4 h at 37°C. It was then washed in three drops of distilled water and negatively stained with 2% sodium phosphotungstate at pH 7·0 for 30 sec. Excess stain was removed with filter paper.

In a control of unadsorbed antiserum a "halo" of protein molecules was found around the heads, tails, tail fibres, baseplates and collars of the phage particles and in a control of normal serum from an unimmunized animal no halo was found. Antibody molecules from sera adsorbed with defective lysates were found attached to recognizable structural features of the phage particles. Seven different classes of antigenic components on the phage surface were demonstrated by this procedure. These were the head protein

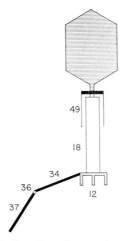

FIG. 1. Demonstration of the location of seven different antigenic components on the surface of phage T4 (after Yanagida and Ahmad-Zadeh, 1970). Specific anti-bodies were found on the head, containing gene 23-protein, the tail sheath (gene 18), the distal tail half-fibre (gene 37), the proximal tail half-fibre (gene 34), the tail fibre joining piece (gene 36), part of the base plate region (gene 12) and part of the head–tail junction (gene 49). Only one of the six tail fibres is shown.

(gene 23), the tail sheath (gene 18), the tail fibre, distal half, (gene 37), the middle part of the two half-fibres (gene 36), the proximal half-fibre (gene 34), part of the base plate (gene 12) and a part of the head-tail junction, the short fibrils attached to the collar (gene 49). Figure 1 shows the location of these parts on the phage particle.

Bradley (1974) used antibodies against bacterial pili to differentiate between the pili and the attached pilus-dependent phages. Carbon-filmed grids were immersed vertically in the bacterial culture for 2 min with gentle agitation and then washed in water. They were then immersed in diluted antiserum for 10 sec, washed in water again and negatively stained in 0·67% neutral phosphotungstate.

Lawn (1967) differentiated between filamentous phage and the bacterial pilus to which it adsorbed by treating the phage-bacterium complex with 0·5% formaldehyde, placing a droplet on a carbon-filmed grid and adding a droplet of suitably diluted rabbit antiserum against the phage. After waiting for 3 min for the antibody to attach the grid was washed with water and stained with 1% uranyl acetate. The filamentous phage particles appeared with a "halo" of adsorbed antibody molecules and were clearly distinguished from the pili to the ends of which they were attached.

Milne and Luisoni (1975) have developed a rapid method for demonstrating specific antibody coating of plant viruses which, in principle, could be applied to any viruses or particles which adsorb strongly to grids. A piece of virus-infected leaf, 2 mm square was crushed with a slender glass rod in 10 μl of water on a glass slide. A carbon-coated grid was touched to the surface of the sample and after a few seconds was rinsed in 30 consecutive drops of 0·1M phosphate buffer, pH 7. The grid was drained but not dried and a drop of antiserum (3–4 μl) was added at a dilution 20 times less than the dilution end point. The grid, still held in forceps, was incubated for 15 min at room temperature in a humid chamber. It was then washed with 20 drops of buffer and then 50 drops of distilled water and finally with 5 drops of 2% uranyl acetate before draining and drying. Some viruses do not survive this treatment, e.g. maize rough dwarf virions, but these can be protected by fixation with 2·5% glutaraldehyde in 0·1M phosphate buffer after the first washing with buffer. The grid was again washed with buffer before proceeding to the antiserum treatment. The antigenic reaction was not impaired. The grids were rendered hydrophilic by placing in a glow discharge at 0·1 Torr for 10 sec.

V. ELECTRON MICROSCOPY OF THE NUCLEIC ACIDS

All the nucleic acids whether DNA or RNA or whether single-stranded or double-stranded are long or extremely long molecules with a very narrow width of about 2 nm. The length of the single piece of DNA isolated from the head of a T-even phage is about 52 μm while the DNA of its host is about 2 mm long. The visualization of molecules of nucleic acid was very difficult and the results had very little useful content until the surface spreading technique was developed by Kleinschmidt and Zahn (1959). The first difficulty was that the molecules did not give sufficient contrast by the negative staining method and, being so narrow they did not give a well defined positively stained image. Shadowing techniques were similarly ineffective in revealing single strands of nucleic acid. A further difficulty was the extreme length of the molecules which easily became tangled together and worse still in the case of single-stranded molecules

they frequently interacted along their lengths to give bunched up structures. Yet another difficulty was that the molecules were easily sheared into
shorter lengths during the preparative procedure and on the grids.

At physiological pH's all nucleic acid molecules are negatively charged
and are neutralized by cations when in the isolated protein-free condition.
This property is the key to the success of the Kleinschmidt method
because it enables the nucleic acid strand to be converted to a thicker
structure by the electrostatic combination with a positively charged protein
of low molecular weight. The protein of choice is almost always cytochrome
c. The thickened filament of nucleic acid can be easily visualized by metal
shadowing. A further feature of this method is that the protein-coated
nucleic acid strands are adsorbed to a film of denatured cytochrome c which
is formed at the surface of a fluid. This film holds the strands firmly and
prevents them becoming tangled or broken. The difference between single
and double strands can be detected by the thickness of the images and the
interaction between regions of single-stranded molecules can be prevented
by making the preparations under conditions where hydrogen bonding is
minimized.

Developments of this technique have enabled the visualization of a great
many nucleic acid molecules isolated from viruses, bacterial plasmids, viral
replicative forms of DNA and work is now progressing to the bacterial
chromosome (e.g. Delius and Worcel, 1973). The method has enabled an
estimate of the molecular weight of DNA to be made independently of
ultracentrifugal or other physical methods. The attachment of specific
proteins such as RNA polymerase and "unwinding" protein can be studied
and the location of deletions and insertions into DNA strands can be found
by heteroduplex analysis. Since all this information may be obtained by the
use of a very small amount of nucleic acid, e.g. 50 μl of a solution containing
0·5 μg/ml, it can be seen that the Kleinschmidt surface spreading technique
is of immense value and is extremely economical in material.

A. General methods

The original surface spreading method of Kleinschmidt and Zahn (1959)
depended on making a mixture of DNA and cytochrome c in an ammonium
acetate buffer and causing a small quantity to run slowly down a ramp onto
the surface of a solution of ammonium acetate of somewhat lower concentration in a special trough. As it spread out over the surface the cytochrome
became denatured at the air–liquid interface and formed an insoluble layer
at the surface. The DNA combined with the cytochrome in solution and
after becoming disentangled, attached itself to the cytochrome film. The
DNA mixture is referred to as the spreading mixture and the fluid on which
it is spread, the hypophase. As it is the intention to adsorb cytochrome to

the DNA strands before they become attached to the denatured cytochrome surface film the success of the method depends on adjusting the ionic conditions so that the interaction between the cytochrome and the DNA is optimum for "thickening" the strands and adsorbing them to the surface film while at the same time avoiding aggregation of the DNA itself. This is achieved by controlling the ionic concentration of the solutions and the concentration of the cytochrome. Another approach called the "diffusion method" has been described by Lang *et al.* (1967). In this the DNA is dissolved in the hypophase and a monolayer of denatured cytochrome is formed on the surface. The DNA molecules, moving about by diffusion pick up cytochrome molecules as they encounter them and eventually reach the surface film to which they firmly adsorb. Again the reaction between nucleic acid and cytochrome is dependent on ionic environment but the amount of nucleic acid attached to the film is a function of time.

The preparation of single-stranded DNA or RNA presents special problems due to the relative thinness of the strands and to their ability to undergo random base–base interactions which result in the strands forming bunches instead of free unentangled filaments. This difficulty has been overcome by the judicious use of denaturing agents, in particular formamide (Westmoreland *et al.*, 1969) which suppress hydrogen bond formation. Davis *et al.* (1971) have referred to the Kleinschmidt method as the "aqueous method" and the Westmoreland method as the "formamide technique".

1. *The aqueous method*

The equipment (Fig. 2) required consists of a trough which can be a glass or plastic Petri dish 10–15 cm diameter or a metal container, ideally coated with Teflon. Kleinschmidt recommends that the glass or plastic troughs should be coated with paraffin wax and should be polished by rubbing with paper before use. Teflon-coated dishes can be cleaned with ethanol and sterilized by heating to 180°C. A Teflon- or wax-coated bar spanning the width of the trough is required on certain occasions. The trough should be 5–10 mm deep and capable of being scrupulously cleaned from dust, finger marks, grease and surface active substances.

The ramp down which the spreading solution is flowed can be a standard microscope slide or a sheet of stainless steel of the same dimensions. It should be cleaned in 1 : 1 nitric acid, if glass, or in alcohol if metal. The glass slide should be washed in distilled water and stored in same until required. The metal ramp should be flamed in a bunsen burner. All ramps should be rinsed in the hypophase fluid before setting at the proper angle in the trough.

The water used for making up the hypophase and the spreading solution should be twice distilled. Kleinschmidt suggests that if the water is sufficiently clean the air bubbles which form when it is shaken must disappear in not more than 1 sec.

Dust particles and any other particulate matter cause clumping of DNA molecules and make length measurement difficult. Westmoreland *et al.* (1969) recommend that the cytochrome solution be filtered through a 0·2 μm filter (Flotronics FM13) and the other solution through Millipore 0·22 μm filters.

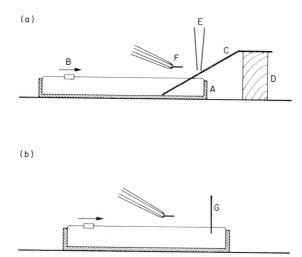

FIG. 2. Preparation of nucleic acid by (a) the surface spreading method and (b) by the diffusion method. In (a) the hypophase is contained in Teflon or wax-lined trough A fitted with a bar B which can be moved (arrow) to compress the surface film. A ramp C of stainless steel or glass is held at an angle in the hypophase by resting on a block D. The spreading mixture containing the DNA, cytochrome and other constituents is run down the ramp from a wide-bore syringe E. The spread film is picked up by touching a coated grid F to the surface. In (b) the DNA is placed at high dilution in the hypophase in the trough. Cytochrome *c* is spread on the surface from a needle G bearing a small amount of the dry protein.

The hypophase for double-stranded DNA is usually 0·25M ammonium acetate at pH 7·5. The spreading solution contains DNA at 0·5–1·0 μg/ml and cytochrome *c* at 100 μg/ml in 0·5M ammonium acetate at pH 7·5. Sometimes EDTA at 1mM is added.

The procedure is to fill the trough with hypophase until the liquid level is slightly above the edge. The surface is then swept clean by passing the bar over it. The ramp is then flushed with hypophase and lowered into the trough. Previous tests should have been made to find a suitable means of resting the upper end of the ramp on a small wooden block, for example, so that the spreading solution runs slowly but steadily down it. The spreading solution is now taken up in a suitable pipette such that 50 μl can be slowly expressed onto the ramp at a distance of about 10 mm from the hypophase surface. A 50 μl pipette of the disposable tip variety can be used provided its action is sufficiently smooth. Whatever pipette is used it should be noted that narrow bore orifices can cause shearing of long DNA molecules and it is therefore necessary to use as wide a bore tip as possible consistent with being able to hold the spreading solution.

It is frequently suggested that the surface of the hypophase be dusted with fine talc particles before the specimen is run down the ramp. This serves no purpose other than to indicate that the protein has spread on the hypophase surface and to give an indication how far it has spread. Too heavy a dusting with talc could prevent the spread of the film and it should be noted that talc of the cosmetic variety is usually treated with perfume, etc.

The film is immediately picked up by touching filmed grids to its surface about one grid width from the ramp-hypophase boundary. It has been noted by Delius et al. (1972) that the properties of the two sides of the grid are not equal in respect of granularity and that better results can be expected if the "air" side of plastic films is applied to the floating protein film (see p. 179). It is important that the grid surface be not hydrophobic and each time a grid is touched to a film it should be examined when lifted to see that a droplet is hanging beneath. If this is not so the grids must be treated as described on p. 184, or fresh grids prepared. The droplet is removed by dipping the grid in 95% ethanol and then removing the ethanol with filter paper. Some authors dip the grids in isopentane after ethanol.

The grids may now be shadowed by the two direction method or better by the rotary method. Platinum, platinum–carbon, platinum–palladium, uranium, uranium oxide and tungsten have been used. The angle of shadowing is low, usually 5°–10°. Kleinschmidt (1968) has used 50 mg of uranium premelted on a tungsten filament and covered with a 50 mg piece of uranium foil at a distance of 15 cm from the rotated specimens. Methods for evaporation of heavy metals are given on p. 185.

It is possible to develop sufficient contrast to photograph the specimen by staining the nucleic acid strands with uranyl acetate (Gordon and Kleinschmidt, 1968). A stock solution of uranyl acetate is made up at 5×10^{-3}M in ethanol and diluted 1 : 50 in acetone just before use. The grid with dependent drop of hypophase is touched to the surface of the stain for a period of 15 sec or longer and then touched to ethanol and dried. If the hypophase contains more than 0·1M ammonium acetate the grid is touched to the surface of water for 2–30 sec before staining.

In another staining procedure (Davis et al. 1971) a stock solution of uranyl acetate at 5×10^{-2} or 5×10^{-3}M in water containing 50mM HCl is made up and stored in the dark. The staining solution is made by diluting the stock to 5×10^{-5}M in 90% ethanol and using it within the hour. Grids are taken straight from the spread film, dipped in the stain for 30 sec and rinsed in isopentane for 10 sec. A typical electron micrograph of double-stranded DNA prepared by the Kleinschmidt method is presented in Fig. 5.

2. Lang's diffusion method

This has been described by Lang et al. (1967) and Lang (1971) (Fig. 2). The DNA is dissolved in 0·2M ammonium acetate at 5×10^{-8} μg/ml and the solution used to fill the trough. The amount of DNA required is very much less than with the surface spreading method. After allowing a few minutes for temperature equilibration the cytochrome is applied by dipping a needle bearing a small amount of the dry protein into the fluid in the trough. The cytochrome dissolves and spreads out as a monolayer on the surface. A period of time from 10–20 min is allowed for the DNA to diffuse to the surface film and attach to it, then grids are applied in the usual way to pick up portions of the film which is then dried and shadowed. Lang (1971) has also described a micro method in which 40 μl droplets of a mixture of DNA (0·1 μg/ml), cytochrome c (1·3 μg/ml) in 0·15M ammonium acetate containing 0·07M formaldehyde are placed on a Teflon-coated surface. After allowing 10 min for the DNA to attach to the film of denatured cytochrome which spontaneously appears at the droplet surface a coated grid is applied to pick up a portion of the film bearing the DNA molecules. The grids are dried and shadowed as before. This method uses a very small amount of nucleic acid.

3. Inman's method

Inman and Schnös (1970) have devised a droplet method (Fig. 3) in which a Teflon block 15 cm × 15 cm × 0·3 cm has 24 hemispherical indentations 1·9 cm diameter and 0·1 cm deep milled on its surface. Drops of double distilled water (1·2 ml) are placed in the indentations and a water-wetted clean glass rod 3 mm diameter with a smooth rounded end is held at an angle just entering the liquid surface. The spreading solution (5 μl) is run from a capillary pipette 1·1 mm internal diameter onto the rod from whence it runs onto the water droplet. The rod is then removed and the droplet reduced in volume by 0·1 ml by a syringe fitted with a fine needle. This compresses the film on the drop surface. The DNA sample is picked up on a carbon film evaporated onto a freshly cleaved mica disk 10 mm diameter. The disk is washed by immersion in ethanol and dried in a stream of warm dry nitrogen gas. It is then rotary shadowed with platinum and the carbon film floated off onto water. The film is picked up on grids. The spreading solution consists of NaCl (0·1M), KH$_2$PO$_4$ (6·7mM), EDTA (3·4mM) and formaldehyde (10%) adjusted to pH 6·5–7·5. DNA is added to an optical density of 0·01–0·005 at 260 nm. This mixture is heated to temperatures in the range 48°C–59°C for 10 min to thermally denature the DNA of the phages P2 and lambda for the purpose of preparing thermal denaturation maps. The partially denatured DNA is then mixed with an equal volume of formamide before spreading (see Formamide Method, below). The carbon-coated disks give best results if they are aged for a number of days by storage at low humidity but after 4 months they no longer give good results.

4. The formamide method

This has been described by Westmoreland *et al.* (1969) and Davis *et al.* (1971). Formamide is a denaturing agent which, under suitable conditions,

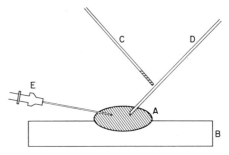

Fig. 3. Inman's method for preparing DNA. A drop of the hypophase A (1·2 ml) is placed in a hollow in the Teflon block B. The spreading solution is run from the wide-bore syringe C onto glass rod D immersed in the drop. After spreading the surface of the drop is slightly compressed by withdrawal of 0·1 ml of the hypophase by syringe E.

can separate the two strands of the DNA double helix. The forces holding together the two strands are principally hydrogen bonds between the complementary bases adenine and thymine and guanine and cytosine. Adenine is held to thymine less strongly than guanine is held to cytosine and it therefore follows that regions of the double helix which are relatively rich in adenine and thymine base pairs can be denatured more easily than other regions. Single-stranded DNA can bind weakly to itself or to other single strand DNA by random base pairing but these bonds are easily broken by mild denaturing agents. Strands of the DNA duplex can also be separated by heat and the temperature at which complete separation occurs is referred to as the melting temperature, T_m (see Deley, this Series, Volume 5A). Formamide in effect lowers the melting temperature by an amount depending on its concentration. The cation concentration has a marked effect on denaturation, the higher the concentration the more formamide is needed to effect the same extent of strand separation. Increase in temperature of course aids strand separation.

The formamide method enables the separated single strands to be visua-

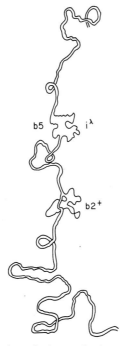

FIG. 4. An interpretative drawing of a heteroduplex made by annealing the *l* strand of phage lambda DNA with the *r* strand of a mutant lanbda *b2b5* in which the *b2*+ region is deleted and the *i* region is replaced with a shorter nonhomologous region *b5*. The heteroduplex is double-stranded throughout its length except for the *b2*+ region which appears as a single-stranded loop and the *b5* and *i* regions, which are also single-stranded, but of unequal length (after Westmoreland *et al.*, 1969). Reproduced by kind permission of the American Association for the Advancement of Science, copyright 1969.

lized and their positions along the duplex to be measured. It has been particularly useful for the physical mapping of the base sequence homology of two DNA's by a procedure known as heteroduplex analysis (Westmoreland *et al.*, 1969) where it was applied to deletion and substitution mutants of phage lambda (Fig. 4). Numerous other studies have been made for example between the related but not identical phages T3 and T7 (Davis and Hyman, 1971).

In the procedure used by Westmoreland *et al.* (1969) for the examination of heteroduplexes of lambda phage DNA the *l* and *r* strands were separated on CsCl gradients. Fractions (10 μl) containing each of the two molecules, but derived from different mutants of the phage, in 6M CsCl were added to 1 ml of 50% formamide made by adding 1 ml of formamide to 1 ml of 0·01M NaHCO$_3$ at pH 8·6. The mixture was stood at 4°C for 5 days or longer to allow the strands to anneal together. The final DNA concentration was 1–4 μg/ml in 0·12M CsCl. The spreading mixture contained 0·1 ml of the DNA solution in 50% formamide and 0·01 ml of 0·1% cytochrome *c* in water. The hypophase was double distilled water. The monolayer was picked up by touching formvar-coated grids to the surface and dehydrating for 5 sec each in absolute ethanol, amyl acetate and 2-methyl butane. The grids were shadowed with uranium oxide. A commercial uranium oxide which contained a nitro-cellulose binder was treated with acetone to remove half the binder. The dried oxide (12 mg) was evaporated from a tungsten basket at an angle of 6°–7° at 11 cm from the rotated specimen grids. The nitrocellulose was first removed by slowly heating at a pressure of less than 0·1 μmHg. The uranium oxide was then evaporated by rapid heating for 5–10 sec at a pressure lower than 0·03 μmHg.

In another procedure the spreading solution consisted of DNA at 0·5 μg/ml in a solution of 0·1M Tris and 10mM EDTA at pH 8·5 in 40% formamide in water. Cytochrome *c* was added at 0·02–0·1 mg/ml. At this salt concentration lambda phage DNA is half denatured when the formamide concentration is 85%. The hypophase consisted of 10mM Tris and 1mM EDTA (pH 8·5) in formamide at a concentration 30% below that in the spreading solution. The hypophase was made up not more than 5 min before use because it may become acid if kept longer. The glass ramp was washed with ammonium acetate (0·2M) and allowed to drain dry before setting in the trough of hypophase. The spreading solution (50 μl) was gently run down the ramp close to the ramp-hypophase boundary. After standing for 1 min the film was picked up and stained in uranyl acetate. The spread film may be compressed by pushing it back with a Teflon bar in order to increase the contrast. Staining is of little use on its own and it is recommended that the grids be shadowed.

5. *Single-stranded DNA*

The DNA of some viruses is entirely single stranded, e.g. bacteriophage ØX 174 and the filamentous phages fd and M13. These DNA's are also circular. Bujard (1970) has examined the DNA of phage fd by the following procedure. The phage was diluted to give 100–400 μg of DNA per ml in a buffer containing 0·01M sodium phosphate and 0·001M EDTA at pH 7·5. Sarkosyl 90 was added to 2·5% and the DNA was extracted with buffer-saturated phenol by gently shaking at room temperature for 20 min. The phenol was removed from the aqueous phase by dialysis against buffer. The DNA was unfolded by adding 0·3 ml containing 10–15 μg to 1·2 ml of a solution composed of 0·05M sodium phosphate, 5% formaldehyde and 26% dimethylsulphoxide at pH 7·6. The mixture was heated at 50°C for 30 min and chilled in ice for 5 min. This DNA was diluted 0·3 ml into 50 ml of 0·2M ammonium acetate and spread by the diffusion method of Lang (see p. 201). The DNA appeared as small circles whose contour length was markedly affected by the ionic strength of the ammonium acetate. The length increased by 74% when the ammonium acetate concentration was decreased from 0·4 to 0·1M. It was concluded that the best preparations of the DNA were obtained at 0·1 and 0·2M where measurements on the length of the molecules showed the least standard deviation. It was also observed that the contrast of the DNA molecules was a maximum when the pH of the mixture was 6·0–6·5. Above 7 the contrast is lower and at pH 8 the strands can hardly be seen. The mean length of fd DNA by this method was 1·25 μm ± 5% (sample S.D.).

6. The examination of RNA

RNA can be prepared by monolayer spreading techniques if steps are taken to prevent random base associations which would result in the strands coiling upon themselves. Granboulan and Scherrer (1969) extracted RNA from HeLa cells with hot phenol, removed the DNA with DNA'se and fractionated the RNA on 15–30% sucrose gradients containing 0·01M EDTA and 0·05M NaCl. The RNA (0·05 ml) was added to 1 ml of 8M urea to give a final concentration of about 1 μg/ml. A solution of cytochrome c (0·01%) in 1M ammonium acetate at pH 8 was filtered through a 0·22 μm Millipore filter and 0·1 or 0·05 ml added to 0·1 or 0·2 ml of the RNA solution in urea. This was allowed to flow down a ramp on to a hypophase of 0·015M ammonium acetate. The film was picked up as for DNA and rotary shadowed with uranium oxide at 5×10^{-5} Torr. Pieces of RNA ranging from 0·4–3·5 μm were found. The 0·4 μm pieces were believed to be 16S RNA. The average weight of sodium ribonucleotide is 343 daltons and the length per base in a polynucleotide has been estimated at 3·03 or 3·14 Å. The length of phage R17 RNA, which is known to have 3242 nucleotides, is 1·06 μm which in this case gives an internucleotide distance of 3·17 Å.

Robberson et al. (1971) have examined the lengths of ribosomal 12S, 16S and 18S RNA isolated from HeLa mitochondrial and cytoplasmic ribosomes. The RNA species were purified by three successive sucrose gradient centrifugations and by polyacrylamide gel electrophoresis to remove traces of DNA. The purified RNA was precipitated with ethanol and redissolved in 10^{-3}M Tris buffer containing $2·5 \times 10^{-4}$M disodium EDTA at pH 8·0 to a

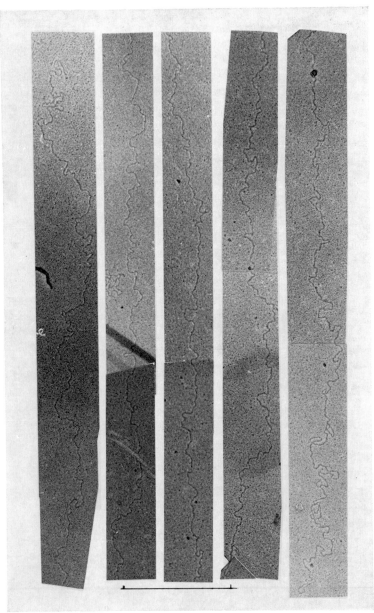

FIG. 5. Double-stranded DNA extracted from bacteriophage T5 and prepared by the Kleinschmidt method. The strand, which is 36 μm long, is a complete phage genome. It runs from the top of the left-hand photograph to the bottom of the right-hand one. Each photograph either overlaps the next or butts exactly to it. (Specimen prepared by R. Everett.) Bar equals 1 μm.

final concentration of 40–50 μg/ml. For electron microscopy 0·5 μl of the sample of RNA was diluted into 100 μl of a 4M solution of urea made up in formamide which had been specially purified by the following procedure. Commercially available 99% formamide in 350 ml quantities in 400 ml beakers was immersed in an ice bath and magnetically stirred. When 40–50% of the formamide had crystallized the bulk of the supernatant fluid was decanted and the crystals centrifuged on sintered funnels specially fitted into the buckets of a centrifuge. The crystals were stored at $-70°C$. The diluted RNA sample was heated to 53°C for 30 sec. After cooling to 25°C about 2 μl of a solution containing 2·5 mg/ml of cytochrome in 2M Tris and 0·05M disodium EDTA were added. The final concentrations were RNA, 0·20–0·25 μg/ml; cytochrome, 50 μg/ml; urea, 4M; Tris buffer, 0·04M; EDTA, 0·001M and formamide, approx. 80% w/v. This solution (35 μl) was spread onto a hypophase (140 ml) of 10^{-6}M Tris, $2·5 \times 10^{-7}$M disodium EDTA (pH 7·8) and 2 μl of diethylpyrocarbonate. The grids were picked up and shadowed as for DNA.

Lengths of 0·27 μm, 0·42 μm and 0·55 μm were obtained for 12S, 16S and 18S ribosomal RNA. If the lengths are taken to be proportional to the molecular weight and the molecular weight of 18S cytoplasmic ribosomal RNA is taken to be $0·71 \times 10^6$ daltons then the molecular weights of the 12S and 16S RNA are $0·35$ and $0·54 \times 10^6$ daltons respectively. While these values are in good agreement with those predicted from sedimentation velocity measurements they do not agree well with those based on gel electrophoresis. From the electron microscope measurements the average internucleotide spacing for ribosomal RNA was 2·6–2·7 Å. For ØX 174 single-stranded DNA the internucleotide spacing is 2·9–3·0 Å in 50% formamide while for double-stranded DNA the values are 3·3–3·5 Å per base pair.

In contrast to the RNA molecules described by Granboulan and Scherrer (1969) those from the influenza virus can be spread either with or without urea and formamide (Li and Seto, 1971). The reason for this difference is not clear but presumably depends on the base sequences and the secondary structure of the RNA. The nucleic acid was extracted from the purified virions of the X7, X7–F1 and WSN strains of influenza virus by the use of sodium dodecyl sulphate and phenol and precipitated with cold ethanol. The RNA was centrifuged down and resuspended in buffer containing 0·1M NaCl, 0·01M Tris-HCl (pH 7·4) and 0·001M EDTA. It was then subjected to sucrose density centrifugation (5–20%) in a buffer one-tenth the above strength. After 4·5 h at 49,000 rpm in a Spinco SW 50L rotor the gradient was fractionated and the RNA immediately spread as follows. The RNA was diluted in buffer to 2–5 μg/ml and 0·2–0·3 ml was added to an equal volume of filtered cytochrome c at 1·0 mg/ml in 0·5M ammonium acetate. The spreading solution was run down a glass ramp onto a hypophase of 0·02M ammonium acetate at pH 8·0 in a sterile plastic Petri dish. The protein film was immediately picked up on grids, rinsed in water for 20 sec and stained in uranyl acetate (10^{-4}M) containing 5×10^{-4}N HCl in ethanol for 30 sec and dehydrated in 2-methylbutanol for 10 sec. The RNA of the X7 strain of virus had a mean length of $2·69 \pm 2·0$ μm and a modal length of 2·7 μm based on 300 measurements. Storage of the RNA at 4°C for a few days resulted in fragmentation of the strands.

The genome of the influenza virus is known to consist of five discrete pieces. These could be demonstrated in the electron microscope after exposing the RNA to a pH of 3 which suppresses hydrogen bonds. After spreading the specimen was found to contain fragments whose length varied from 0·18 to 0·96 μm. Five distinct peaks could be seen in a histogram with mean lengths of 0·27, 0·42, 0·59, 0·74 and 0·88 μm. Li and Seto (1971) used the figure of 345 daltons for the average sodium ribonucleotide in influenza virus RNA and the figure of 3·17 Å for the base spacing as in phage R17 RNA. These figures gave molecular weights of 2·9, 2·8 and $2·5 \times 10^6$ daltons for the X7, X7–F1 and WSN viral RNA respectively.

The protein monolayer technique has been extended to plant cell viroid by Sogo *et al.* (1973). Potato spindle tuber viroid (Diener, 1971) is an infectious RNA which is too small to contain genetic information for self replication and is devoid of protein capsid. The spreading solution was composed of viroid in water at 30 μg/ml diluted to 2–10 μg/ml in 4M sodium acetate (pH 7·3) to which an equal volume of 0·05% cytochrome *c* in 4M sodium acetate, freshly filtered through a 25 nm Millipore filter had been added. This was spread at room temperature on a hypophase of twice distilled water. The films were picked up on carbon films evaporated onto mica. These were floated off onto water and allowed to settle onto 400 mesh grids. They were washed for 10–15 min in water, stained for 30 sec in 2% methanolic uranyl acetate, washed for 10 sec twice in methanol, dried and shadowed with platinum. In a second procedure the spreading solution was made by adding 0·05 ml of the RNA dissolved in 0·1M NaCl and 0·01M EDTA to 1 ml of freshly filtered (25 nm) 8M urea. This was diluted 1 : 1 or 1 : 3 with 0·02% cytochrome *c* in 1M ammonium acetate (pH 8·0) to give final viroid concentrations of 0·6 and 0·3 μg/ml. The hypophase was 0·015M ammonium acetate at pH 8 or distilled water. Very short strands with a mean length of $51·2 \pm 4·1$ nm were found. These were absent when the viroid was pretreated with RNA'se. The molecular weight was estimated to be $7·9 \times 10^4$ daltons.

B. Nucleic acid molecules with attached proteins

The Kleinschmidt procedure and its variants so far described have dealt solely with naked nucleic acid strands. Frequently these have been isolated from cells or virus particles by procedures which strip away any proteins which might have been associated with the nucleic acids. These proteins would include histones, polymerases, unwinding proteins and proteins which link nucleic acid strands together. In order to study these proteins and their interactions with nucleic acids methods are needed which enable the strands to be isolated with the proteins attached and preparative methods are required which enable the protein-bearing strands to be visualized in the electron microscope. It is also possible to prepare complexes between nucleic acids and specific proteins *in vitro* and methods are required to examine their structure. Examples are given below in which the DNA specific protein produced under the control of phage T4 gene 32 can be visualized attached to phage single stranded DNA. In another example the complexes between SV 40 viral DNA and host cell histones are examined.

1. *Gene 32-protein and phage DNA*

The protein specified by T4 gene 32 is able to bind tightly and co-operatively to single stranded DNA. It is an essential protein for phage DNA replication in T4-infected bacteria and for genetic recombination in these cells. The protein is able to invade adenine-thymine rich double helical regions without the requirement for a physical discontinuity such as a cleaved phosphodiester bond or a double-stranded end. The protein has been proposed by Delius *et al.* (1972) as a cytological tool for electron microscopical characterization of DNA structure and has been used in such investigations, e.g. Monjardino and James (1975) in a study of polyoma virus DNA.

The gene 32-protein was prepared by the method of Alberts and Frey (1970) and purified on a hydroxyapatite column. The reaction with single-stranded DNA was demonstrated by taking filamentous phage fd DNA (a single-stranded closed circle) at 1 μg/ml and incubating it with 30 μg/ml of gene 32-protein in 0·01M potassium phosphate buffer at pH 7·0 for 10 min at 37°C. The complex formed is unstable in the presence of formamide used to spread it on the hypophase. It was rendered stable by treatment with glutaraldehyde at 0·01M for 15 min at 37°C followed by dialysis against 0·1M Tris-HCl at pH 8·5. A spreading solution was then made with 30% formamide and 0·01% cytochrome *c*, the DNA being diluted to 0·1–0·5 μg/ml with Tris buffer. The formamide was flash distilled and purified as described on p. 205. Under these conditions the complexes appeared as uniform circles of protein covered DNA with a thickness slightly greater than observed for double-stranded DNA. The uncomplexed single-stranded DNA from phage fd had a length of 1·92 ± 0·15 μm but after attachment of the gene 32-protein there was an increase of length by about 50% to 3.02 ± 0.15 μm.

Reaction of gene 32-protein with DNA isolated from phage lambda produced a series of short bifurcations in the otherwise linear native DNA structure. The single strands in these loops appeared thicker than the double-stranded DNA thus demonstrating that they were coated with the gene 32-protein. This result was obtained with 10 μg/ml of DNA and 100 μg/ml of gene 32-protein but when the protein concentration was increased to 200 μg/ml the DNA molecules became extensively denatured. A denaturation map was made using 10 μg/ml for 10 min at 30°C and pH 7 and found to be very similar to the alkaline denaturation map obtained by the method of Inman and Schnös (1970). This indicates that the gene 32-protein preferentially invades the same adenine-thymine rich regions which are sensitive to mild alkaline denaturation.

Complexes between phage PM2 DNA, which has a molecular weight of 6.3×10^6 daltons and is very tightly supercoiled (Espejo et al., 1969), and gene 32-protein have been made by Brack et al. (1975) by the procedure of Delius et al. (1972) described above. The complexes were fixed in 0.1% glutaraldehyde for 10 min at 37°C and purified by gel filtration (Biogel A, 1.5M), eluting with 1mM EDTA. The complexes were prepared for electron microscopy by allowing them to adsorb to positively charged grids made by the procedure of Dubochet et al. (1971) (see p. 184). The grids were stained in 2% aqueous uranyl acetate for 2–3 sec and shadowed with platinum. The nucleic acid was found to contain one to three single stranded regions detectable by their reaction with gene 32-protein or by alkali denaturation. These regions were found to occur in any of eight different positions in the genome which could be mapped by reference to the unique site at which the restriction endonuclease R.*Hap* II cleaved the molecule.

2. *SV 40 virus minichromosomes*

When permissive monkey cell cultures are infected with SV 40 virus a pool of circular viral molecules accumulates in the cell nuclei. The viral DNA can be isolated as a complex of almost equal weights of DNA and cellular histones. When these complexes are prepared in the presence of 0.15M NaCl and sodium bisulphite to inhibit histone proteases they can be visualized in the electron microscope as small closed loops of a fibre 11 nm diameter and 210 nm long. These were termed "minichromosomes" (Griffith, 1975). The grids were prepared by the method of Griffith (1973) (see p. 184). The samples were adsorbed to the grids, after fixation in 0.1% glutaraldehyde at 20°C for 15 min, washed in water, dehydrated in ethanol and "stained" by the tungsten evaporation method (see p. 187). If the "minichromosomes" were diluted ten-fold in distilled water before fixation they appeared in an altered form consisting of 21 beads about 11 nm diameter connected together in a ring by a thin thread about 2 nm diameter.

C. Determination of molecular weights of DNA

The determination of the molecular weight of DNA by electron microscopy employs the relationship between the apparent contour length L and the molecular weight M (Lang, 1970).

$$M = M'L \text{ (daltons)}$$

The contour length is found by measuring the images of the DNA molecules as prepared by a standard method and M' (daltons/cm) is the molar linear density which is obtained by calibration with DNA of independently determined molecular weight. The conditions used and the manipulations involved in preparing specimens for electron microscopy can cause distortion of the molecular configuration of the DNA and result in values for the contour length which vary from -50% to $+100\%$ of the true value in extreme cases. The factors which can cause distortion of nucleic acid molecules have been listed by Lang (1970) and include ionic strength of the DNA solution, binding to the cytochrome c and adsorption to the interface, mechanical instability of the protein film, adsorption to the specimen grids, drying in ethanol and shrinkage or expansion of the specimen supports in the electron beam. It has been shown by Lang and Mitani (1970) that DNA with two nicks on opposite strands may break even if the nicks are 20 nucleotides apart. However by the use of a standardized procedure results which are reproducible to a few per cent can be obtained. The method has the advantage of being able to give an independent estimate of the molecular weight of DNA by the use of a very small amount of material.

Lang (1970) determined the relative molecular weights of the DNA from bacteriophages T4, T5 and T7 and by reference to the independently determined molecular weights of those nucleic acids (Freifelder, 1970, in a companion paper) was able to obtain a value for M' and to give lengths for each nucleic acid. The bacteriophages were grown as described by Davison and Freifelder (1962) purified by four cycles of low and high speed centrifugation followed by two cycles of isopycnic centrifugation in CsCl containing $0.01M$ $MgCl_2$ and in the case of phage T5, $10^{-3}M$ $CaCl_2$. The phages were dialysed four times against $0.01M$ Tris (pH 7.8), $0.1M$ NaCl, $0.01M$ $MgCl_2$ and for T5, $10^{-3}M$ $CaCl_2$. The optical densities of the phages at 260 nm were 12 for T4 and T5 and 36 for T7. The DNA was isolated by adding 0.01 ml of stock phage to 0.1 ml of $5M$ $NaClO_4$ at pH 7.8.

The diffusion method was used to prepare the phage for electron microscopy. To 0.05 ml of the released DNA mixture was added 30 ml of $0.20M$

ammonium acetate containing 10^{-3}M EDTA at pH 6·5. The solution was poured into a Teflon-coated dish holding 29 ml (Lang and Mitani, 1970) and left for 20 min for temperature equilibration. The surface was cleaned by drawing a Teflon-coated bar across it. The cytochrome c was spread on the surface by dipping a needle coated with a dry film of the protein into the DNA solution. The ionic strength of the DNA solution was 0·2. This was found to give the most reproducible length for the DNA molecules.

After allowing 15 min for diffusion, portions of the surface film with adsorbed DNA were picked up on carbon-coated grids (in this case Siemens 7-hole type), dried in ethanol and shadowed with platinum. Micrographs were taken at nominal magnifications of 5000 and 10,000 after the microscope lens currents and high tension had been switched on for 30 min. In order to avoid errors in magnification which arise due to differences in specimen position relative to the objective lens in different parts of the same grid and from grid to grid the images were focused by adjusting the vertical position of the grids while keeping the objective lens current constant. The error introduced by variations in specimen position have been estimated by Lang et al. (1967) to be ± 0·8% sample standard deviation. The actual magnification of the image on the plates was determined by taking micrographs of carbon replicas of a cross-ruled grating with 54,800 lines per inch. The accuracy of the grating was itself checked by light microscopy.

The plates were optically enlarged 23·7 times and the image traced onto paper. The length of each molecule was then measured with a map measurer and corrected for pin cushion distortion which was on average 0·5%. A large number of DNA molecules were measured and the lengths obtained together with the molecular weights taken from the companion paper by Freifelder are given in Table 1.

A number of assumptions were made in the calculation of the molecular weights. The mean molecular weight of T7 DNA at $25·1 \times 10^6$ daltons was

TABLE I

DNA	Length (μm)	Total error (± μm)	Molecular weight[†] (daltons × 10^{-6})
BPV*	2·49	0·05	5·15
T7	12·15	0·25	25·1
T5	36·2	0·8	74·9
T4	52	1·0	119

* Bovine papilloma virus.
† Relative to T7 DNA at $25·1 \times 10^6$ daltons.

taken to be correct as it was based on three independent observations. M' was assumed to be practically independent of the base composition and where cytosine is replaced by hydroxymethyl cytosine, as in the T-even phage DNA, this is assumed to have no effect on the DNA configuration. The "B" configuration is assumed to hold for the DNA in solution but there is no certainty that this is true for DNA on the grids. The length per pair of complementary nucleotides is taken to be 3·46 Å. Taking these figures into consideration a molar linear density (M') for T7 DNA of $2·07 \pm 0·04 \times 10^{10}$ daltons/cm was calculated and is believed to be applicable to DNA's with common bases. For T4 DNA M' is $2·28 \times 10^{10}$ and for T2 DNA $2·23 \times 10^{10}$ daltons/cm.

VI. MEASUREMENT PROCEDURES

In order to obtain length measurements with as high an accuracy as possible it is necessary that the magnification of the electron microscope be determined. This is best done with a carbon replica of a cross-ruled diffraction grating and in order to minimize hysteresis effects in the electromagnetic lenses the manufacturer's instructions should be carried out each time the instrument is used. If possible an internal check on the magnification should be made by including some DNA of known length for example one of those in Table 1 or ØX 174 RF DNA, MW $3·6 \times 10^6$ daltons, length $1·63 \pm 0·08$ μm.

Only molecules which do not overlap other molecules or parts of themselves should be measured. Negatives should be magnified 10–50 times in a suitable optical projector, e.g. Nikon shadowgraph, and traced onto paper. The number of molecules to be measured depends on the accuracy required. Davis *et al.* (1971) suggest that 20–100 should be measured for heteroduplex analysis. If a large number of molecules have to be measured and the data accumulated and analysed instrumental aids can be used with advantage. Inman (1974) has used a Hewlett-Packard 9810 or 9820 calculator interfaced with a digitizer and plotter by the same maker. Histograms representing position, size and frequency of denatured sites can be plotted with the aid of a computer. Schnös and Inman (1970) have described the use of a curve length integrative device (Sperry Gyroscope Co.) which automatically computes the distance through which a hand-held pen is moved. The total length of molecules or individual positions of denatured sites and or branches can be determined by following the projected images with the pen.

REFERENCES

Alberts, B., and Frey, L. (1970). *Nature*, **227**, 1313–1318.
Brack, C., Bickle, T. A., and Yuan, R. (1975). *J. Mol. Biol.*, **96**, 693–702.

Bradley, D. E. (1959). *Brit. J. Appl. Phys.*, **10**, 198 *et seq.*

Bradley, D. E. (1961). *Virology*, **15**, 203–205.

Bradley, D. E. (1965). *In* "Techniques for Electron Microscopy", p. 73 (Ed. D. Kay). Blackwell Scientific Publications, Oxford.

Bujard, H. (1970). *J. Mol. Biol.*, **49**, 125–137.

Davis, R. W., and Hyman, R. W. (1971). *J. Mol. Biol.*, **62**, 287–301.

Davis, R. W., Simon, M., and Davidson, N. (1971). *In* "Methods in Enzymology", Vol. XXI, Part D, pp. 413–428 (Eds. L. Grossman and K. Moldave). Academic Press, New York.

Davison, P. F., and Freifelder, D. (1962). *J. Mol. Biol.*, **5**, 635–642.

Delius, H. and Worcel, A. (1973). *J. Mol. Biol.*, **82**, 107–109.

Delius, H., Mantell, N. J., and Alberts, B. (1972). *J. Mol. Biol.*, **67**, 341–350.

Diener, T. O. (1971). *Virology*, **45**, 411–428.

Doyle, B. B., Hukins, D. W. L., Hulmes, D. J. S., Miller, A., and Woodhead-Galloway, J. (1975). *J. Mol. Biol.*, **91**, 79–99.

Dubochet, J., Ducommun, M., Zollinger, M., and Kellenberger, E. (1971). *J. Ultrastruct. Res.*, **35**, 147–167.

Espejo, R. T., Canelo, E. S., and Sinsheimer, R. C. (1969). *Proc. Nat. Acad. Sci. U.S.A.*, **63**, 1154–1168.

Fernandez-Moran, H., Van Bruggen, E. F. J., and Ohtsuki, M. (1966). *J. Mol. Biol.*, **16**, 191–207.

Freifelder, D. (1970). *J. Mol. Biol.*, **54**, 557–565.

Fukami, A., and Adachi, K. (1965). *J. Electron Microscopy*, **14**, 112–118.

Gordon, C. N., and Kleinschmidt, A. K. (1968). *Biochem. Biophys. Acta*, **155**, 305–307.

Granboulan, N., and Scherrer, K. (1969). *Europ. J. Biochem.*, **9**, 1–20.

Griffith, J. D. (1973). *In* "Methods in Cell Physiology", Vol. 7, pp. 129–146 (Ed. D. Prescott). Academic Press, New York.

Griffith, J. D. (1975). *Science*, **187**, 1202–1203.

Harris, W. J. (1962). *Nature*, **196**, 499–500.

Hinnen, R., Schafer, R., and Franklin, R. M. (1974). *Europ. J. Biochem.*, **50**, 1–4.

Horne, R. W., and Nagington, J. (1959). *J. Mol. Biol.*, **1**, 333.

Horne, R. W., and Pasquali Ronchetti, I. (1974). *J. Ultrastruct. Res.*, **47**, 361–383.

Horne, R. W., Hobart, J. M., and Pasquali Ronchetti, I. (1975). *Micron*, **5**, 233–261.

Inman, R. B. (1974). *In* "Methods in Enzymology", Vol. XXI, Part E, pp. 451–458. (Eds. L. Grossman and K. Moldave).

Inman, R. B., and Schnös, J. (1970). *J. Mol. Biol.*, **49**, 93–98.

Johansen, B. V. (1974). *Micron*, **5**, 209–221.

Kleinschmidt, A. K. (1968). *In* "Methods in Enzymology", Vol. XII, Part B, pp. 361–377 (Eds. S. P. Colowick and N. O. Kaplan).

Kleinschmidt, A. K., and Zahn, R. K. (1959). *Z. Naturforsch.*, **14b**, 730.

Kirkpatrick, J. B. (1969). *Science*, **163**, 187–188.

Lang, D. (1970). *J. Mol. Biol.*, **54**, 557–565.

Lang, D. (1971). *Phil. Trans. Roy. Soc. Lond.* B, **261**, 151–158.

Lang, D., and Mitani, M. (1970). *Biopolymers*, **9**, 373–379.

Lang, D., Bujard, H., Wolff, B., and Russell, D. (1967). *J. Mol. Biol.*, **23**, 163–181.

Lawn, A. M. (1967). *Nature*, **214**, 1151–1152.

Li, K. K., and Seto, J. T. (1971). *J. Virol.*, **7**, 524–530.

Louie, D. D., Kaplan, N. O., and McLean, J. D. (1972). *J. Mol. Biol.*, **70**, 651–664.

Mellema, J. E., Van Bruggen, E. F. J., and Gruber, M. (1968). *J. Mol. Biol.*, **31**, 75–82.
Milne, R. G., and Luisoni, F. (1975). *Virology*, **68**, 270–274.
Monjardino, J., and James, A. W. (1975). *Nature*, **255**, 249–251.
Nermut, M. V. (1972). *J. Microsc.*, **16**, 351–362.
Nermut, M. V. (1975). *Virology*, **65**, 480–495.
Rich, A., Warner, J. R., and Goodman, H. M. (1963). *Cold Spring Harbor Symp. Quart. Biol.*, **28**, 269–285.
Schnös, J., and Inman, R. B. (1970). *J. Mol. Biol.*, **51**, 61–73.
Sogo, J. M., Koller, T., and Diener, T. O. (1973). *Virology*, **55**, 70–80.
Robberson, D., Aloni, Y., Attardi, G., and Davidson, N. (1971). *J. Mol. Biol.*, **70**, 473–484.
Wells, B. (1974). *Micron*, **5**, 79–81.
Westmoreland, B., Szybalski, W., and Ris, H. (1969). *Science*, **163**, 1343–1348.
Yanagida, M., and Ahmad-Zadeh, C. (1970). *J. Mol. Biol.*, **51**, 411–421.

CHAPTER VI

Bdellovibrio Methodology

Mortimer P. Starr

Department of Bacteriology, University of California, Davis, California 95616,
U.S.A.

AND

Heinz Stolp

Lehrstuhl für Mikrobiologie, Universität Bayreuth, 8580 Bayreuth,
Federal Republic of Germany

I. INTRODUCTION

Bdellovibrio Stolp and Starr 1963 is a genus of bacteria which have several unusual properties of direct relevance to the methodologies suitable for their study:

A. The bdellovibrios are "symbiosis-competent" bacteria (see Section II for an explanation of this term); that is, they have an unprecedented symbiotic capacity. The small, highly motile, vibrioid *Bdellovibrio* swarmers can enter through the cell wall into the periplasmic spaces of other ("competent" or "congenial") bacterial cells; they live and develop therein; they consume and metabolize—in an exquisitely regulated and highly efficient manner—most of the substance of their associants; they grow (in this periplasmic locus) into non-motile, large, helical or serpentine forms; and, after multiple fission of the helical element and generation of one sheathed flagellum per progeny cell, the several progeny swarmers swim away from the ghosted remnants of the other bacterial cell and each swarmer eventually initiates another such dimorphic life-cycle when a suitable bacterial associant is encountered.

B. On lawns of suitable associant bacteria, symbiosis-competent bdellovibrios form plaques which (unlike the situation with bacteriophages) increase in size over several days.

C. Liquid cell suspensions of suitable associant bacteria, when acted upon by *Bdellovibrio* swarmers, are reduced in turbidity—although very large populations of bdellovibrios are produced, the numbers of the associant bacteria are reduced almost to zero.

D. Some symbiosis-competent *Bdellovibrio* clones have been grown, even to the extent of going through the dimorphic life cycle, on various complex noncellular substances prepared from microbial cells.

E. From symbiosis-competent clones, nonsymbiotic clones can be isolated; the latter can be cultivated (saprotrophically) on various culture media in the complete absence of symbiont cells or their complex products. Such nonsymbiotic bdellovibrios have not yet been isolated from natural habitats. Symbiosis-competent clones (revertants) can be selected from the nonsymbiotic clones.

F. The dimorphic life-cycle already alluded to in connexion with the symbiosis-competent *Bdellovibrio* strains (that is, the alternation of rather small, very actively motile, monotrichous, vibrioid swarmers and quite large, helical or serpentine, nonmotile elements which divide by multiple fission) has also been shown to occur in some axenic cultures of nonsymbiotic derivatives. This dimorphic life-cycle, thus, may well be a general (and quite unusual) feature of all *Bdellovibrio* clones.

A considerable literature has been produced about the bdellovibrios since they were first reported by Stolp and Petzold in 1962. For the moment, it will suffice to refer the reader to various secondary literature sources, a rather extensive series of review articles and monographs (Burnham and Robinson, 1974; Finance *et al.*, 1974; Maugeri, 1969; Ottow, 1969; Shilo, 1969, 1973a, 1973b; Starr, 1975b; Starr and Huang, 1972; Starr and Seidler, 1971; Stolp, 1968, 1973; Varon, 1974). The relevant primary literature will be cited later, in context.

We deal here, rather didactically, with an assortment of procedures and facts pertaining to the habitats, enrichment, isolation, enumeration, characterization, and classification of bdellovibrios. Several subjects which are tangentially related to these topics will be discussed; others, albeit interesting, receive but a few words and a literature citation; still others, with minimal relevance to *Bdellovibrio* methodology are treated with silence, merely because of the requested scope of this presentation—facts in these areas can be traced via the aforementioned review articles and the burgeoning current literature (some of which is mentioned in Section VIII).

II. CONCEPTS AND TERMINOLOGY

A. Organismic associations, in general

It may seem strange to some that we should begin this primarily expository essay on *Bdellovibrio* methodology by expressing concern about concepts and terminology. However, we are quite convinced that the methods used for studying *Bdellovibrio* as well as some of the views held about its association with other bacteria have been coloured by the terminology we (Stolp and Starr, 1963) initially applied: "parasitic, predatory, and bacteriolytic". Although we did not state so explicitly then, what we meant to imply by the term "parasitic" was the close physical apposition of the two bacteria plus the (nutritional) dependence of the *Bdellovibrio* on the other microbe; by "predatory" we had in mind the bdellovibrio's seeming "hunt" for a suitable "prey", its physical "attack" on the other organism, and its eventual consumption of the attacked bacterium; and by "bacteriolytic" we were referring to the overall mechanism by which the *Bdellovibrio* lyses the other microbe. We have scrutinized in some detail the general phenomena of bacteriolysis (Stolp and Starr, 1965) in an effort to clarify the bacteriolytic mechanisms in this system. But the other two terms ("parasitic" and "predatory") have, in our opinion, remained a stultifying influence.

Whether *Bdellovibrio* is a "parasite" or a "predator" or both or neither has been a quarrelsome and confusing element in the research and writing of practically all workers with this fascinating microbe. This ill-focused quarrel, as documented elsewhere (Starr, 1975b), reduces to the statements "that by some definitions and usages of the term 'parasite' *Bdellovibrio* is a parasite and by others it is not a parasite [and that] by some definitions and usages of the term 'predator' *Bdellovibrio* is a predator and by others it is not a predator". The term "parasitoid", used by some entomologists, comes closest to fitting the situation of *Bdellovibrio* and its bacterial associant, but it seems unwise to muddy the situation further now by adopting this parochial entomological term.

We will, therefore, avoid here the use of the "parasite/host" and "predator/prey" terminology. Instead, we will treat the association of *Bdellovibrio* with other bacteria in terms of a generalized scheme for classifying organismic associations. This scheme, based on a broad analysis of many kinds of associations between organisms of all sorts, has elsewhere been sketched out (Starr, 1975a) and applied to the *Bdellovibrio* system (Starr, 1975b). This analysis and other considerations have led to the conclusion that many organismic associations can be covered quite adequately by the term "symbiosis" in the sense of de Bary (1879), as emended slightly, explicated, and updated (Starr, 1975a) to yield the summarizing statement: "a symbiosis is an association between two organisms; the term 'organism' is taken very broadly to mean any cellular or non-cellular biological entity which has the capacities to replicate itself and to respond to evolutionary forces or which, minimally, can commandeer in a self-perpetuating and evolutionarily responsive manner the genetic apparatus and services of another biological entity; the association in a symbiosis must be significant to the well-being (or 'unwell-being' or both) of at least one of the associants; the associants in a symbiosis may be of different taxa or of the same taxon".

Many biologists, including students of *Bdellovibrio*, shy away from the term "symbiosis", perhaps with ample justification in view of the historical confusion over its use (Starr, 1975a). If the word proves troublesome, the reader could mentally substitute "organismic association" (or simply "association") where it appears. In fact, provided the terms are defined rigorously and explicitly and then applied consistently within a given writing, any terminology (even "parasite/host" and "predator/prey") could be used—but there would always be the real danger, in such usage, that the reader might slip into applying "his" meanings which may be quite different from those intended and stated by the writer.

Returning to the classificatory scheme (Starr, 1975a), the significant properties of a particular organismic association are specified therein in terms of a set of criterional continua. Each continuum in the set focuses on a particular aspect of organismic associations. The properties of a particular organismic association are then specified—with whatever precision is warranted by the fact-base and the purpose—on the bases of their qualitative and quantitative positions within each continuum. For further details, see the Table 1 and the text of Starr (1975a).

B. Terminology pertaining to *Bdellovibrio*

1. *Symbiosis-competent bdellovibrios*

A *Bdellovibrio* able to enter into a symbiotic association with other bacteria is herein termed "symbiosis-competent" ("S-C"; conventionally, "parasitic" or "predatory" or "host-dependent"). This overall capacity of *Bdellovibrio* to enter into significant associations with other bacteria involves two intertwined elements that are frequently confused: a physical format element which reflects the relative spatial arrangements of the two bacteria, in which the *Bdellovibrio* either enters into and lives within other bacterial cells or sometimes lives in locations only partly within other bacterial cells; and a physiological element which reflects the dependence (and the effects and modal aspects thereof) of the *Bdellovibrio* cells upon the bacteria with which they are associating. For the physical format element, use of the terms "inhabitational symbiosis" or "exhabitational symbiosis", with specific modifiers (for example, "intramural" or "intraperiplasmic"), has been recommended (Starr, 1975a, 1975b). The physiological element (dependence, effects, and modality element) is dealt with by an array of explicit terms (Starr, 1975a, 1975b).

The ambiguous term "host-dependent" ("H-D") could refer to two quite different situations: the dependence of the *Bdellovibrio* strain upon another organism (the "symbiont" or "associant" or "substrate organism"; conventionally, the "host" or the "prey") either while actually associating symbiotically or while living nonsymbiotically (axenically) but still dependent upon the dead carcasses, cell-free extracts, or other complex derivatives of another bacterial organism. The latter case is of course not symbiotic, but one in which the *Bdellovibrio* is nevertheless dependent for its axenic cultivation upon the complex substances of a potential symbiont.

2. *Nonsymbiotic bdellovibrios*

Nonsymbiotic ("N-S") aspects of *Bdellovibrio* life are of several sorts; the line between them and the symbiotic aspects is by no means clear (Starr, 1975b). There is the situation alluded to in the previous paragraph in which S-C bdellovibrios are being cultivated nonsymbiotically (axenically) in bacterial cell suspensions rendered non-viable (that is, incapable of dividing) by various physical or chemical treatments (Crothers *et al.*, 1972; Huang and Starr, 1973b; Ross *et al.*, 1974) or in complex preparations derived from former symbionts or other bacteria (Horowitz *et al.*, 1974). There might possibly be permanently nonsymbiotic ("symbiosis-incompetent") bdellovibrios somewhere in Nature, but none of these is presently known, perhaps in part because of defects in the concepts and methods hitherto used. Essentially all N-S *Bdellovibrio* clones known to date have been isolated in the laboratory from S-C clones, in which N-S cells seem to occur naturally at frequencies of 10^{-6} or 10^{-7} per S-C cell; a few have been derived from facultatively symbiotic (F-S) strains (see the next paragraph). Such N-S strains have been variously termed "saprophytic" (Stolp and Petzold, 1962) and "host-independent" ("H-I"; Seidler and Starr, 1969a). These N-S bdellovibrios can be cultivated in ordinary, complex laboratory media (such as peptone and yeast extract). When N-S cultures are placed in contact with suitable associant cells, S-C clones are again selected. The term "nonsymbiotic" (N-S), will be used here for all such cases, unless there is some historical reason for using the "saprophytic" or "H-I" terminology.

3. *Facultatively symbiotic bdellovibrios*

There are *Bdellovibrio* strains which are termed "facultatively symbiotic" ("F-S"; conventionally, "facultatively parasitic" or "F-P"). Such F-S strains form plaques on lawns and, with fair quantitative equivalence, also grow as colonies on ordinary symbiont-free nutrient agar media (Shilo and Bruff, 1965; Diedrich *et al.*, 1970; Uematsu *et al.*, 1971). Shilo (1973b) and Varon *et al.* (1974) have suggested that the F-S condition may be general for all *Bdellovibrio* strains.

The foregoing classes (S-C, N-S and F-S) can usually be defined, or at least distinguished, operationally—in terms of their capacities to form (or not form) plaques on lawns of suitable associant bacteria, to grow (or not grow) as colonies on simple nutrient agar media, and to yield derivative clones belonging to another class—as follows:

Class	Plaques on lawns	Colonies on nutrient agar	Derivatives known
S-C	Formed	Not formed	N-S
N-S	Not formed	Formed	S-C
F-S	Formed	Formed	N-S

There are indications that derivatives other than those listed may also arise.

III. HABITATS OF *BDELLOVIBRIO*

A. Occurrence of *Bdellovibrio* in Nature

Symbiosis-competent bdellovibrios have been reported from a variety of habitats. Since the bdellovibrios are strict aerobes, they probably occur in many aerobic environments that support the growth of suitable associant bacteria; certainly, we would not expect to find them in permanently anaerobic environments. Most frequently, they have been found in soil and sewage (Dias and Bhat, 1965; Keya and Alexander, 1975; Klein and Casida, 1967; Mishustin and Nikitina, 1970; Parker and Grove, 1970; Stolp and Petzold, 1962; Stolp and Starr, 1963). They have also been reported quite often to occur in rivers and other bodies of freshwater (Guélin and Cabioch, 1970; Guélin and Lamblin, 1966; Guélin *et al.*, 1967, 1969a, b, c) and in marine environments (Daniel, 1969; Mitchell and Morris, 1969; Mitchell *et al.*, 1967; Shilo, 1966; Taylor *et al.*, 1974). Although there are high concentrations of bacteria in infectious diseases and bdellovibrios do attack plant and animal pathogens under laboratory conditions, there has been to our knowledge no study of the occurrence of *Bdellovibrio* in actual cases of infectious disease in either plants or animals.

B. Symbiont specificity

The first studies on *Bdellovibrio* (Stolp and Petzold, 1962; Stolp and Starr, 1963) reported that bdellovibrios are limited to Gram-negative bacteria as substrate organisms and that the symbiont range (conventionally, "host range") varies depending upon the *Bdellovibrio* strain and the assay method used for determining symbiont specificity. Although a great deal has been published about *Bdellovibrio* since those first studies, knowledge today leads to much the same conclusions (summarized by Starr, 1975b; Starr and Huang, 1972; Stolp, 1973). Some *Bdellovibrio* strains have been reported to be limited to a single symbiont strain, others will enter into associations with many different bacteria belonging to one or several species or genera. In some cases, the reported symbiont range is indeed puzzling in that the attacked and the unattacked bacteria are so closely related that they are practically indistinguishable by the usual bacteriological procedures. The known S-C *Bdellovibrio* strains enter into associations only with some, but not all, kinds of Gram-negative bacteria. No bdellovibrios have yet been found which attack with certainty any Gram-positive bacteria nor any blue–green bacteria (Cyanophyta; Cyanobacteria). There have been no reports of the ability of bdellovibrios to attack any eukaryotic organisms, with the sole exception of a so-called "*Bdellovibrio chlorellavorus*" (Gromov and Mamkaeva, 1972); however, from what is presently known (D. M. Coder and M. P. Starr, unpublished observations), this undeniably interesting, antagonistic, bacterial ectosymbiont of the green alga *Chlorella vulgaris* is not a member of the genus *Bdellovibrio* as presently conceived.

The basis, and indeed the full extent, of this symbiont specificity remain largely unknown. One possible basis is evident in a pioneering study (Varon and Shilo, 1969) in which mutant strains of *Salmonella* and

Escherichia with various defects in the synthesis of cell wall lipopoly-saccharides were used as prospective symbionts. The bdellovibrios attached more efficiently to rough strains having a complete lipopoly-saccharide core but lacking the type-specific O antigens than to the smooth wild type strains (which have the type-specific O antigens) or to the extremely rough strains (which have defective lipopolysaccharide cores). Despite the differences in attachment efficiency, the bdellovibrios developed equally well in clones deficient and adequate in what are assumed to be attachment sites. Other studies along these lines, but with less clear-cut results, are summarized in Starr and Huang (1972). One other case of a known basis for symbiont specificity is that in *Spirillum serpens*, where the outermost structured (hexagonally-packed) layer of its cell wall (Buckmire and Murray, 1970, 1973) appears to protect the *Spirillum* cells from attack by *Bdellovibrio* (Buckmire, 1971). Strains of *S. serpens* with defects in this layer are susceptible, and those with a complete layer are resistant, to *Bdellovibrio*.

Since the apparent symbiont specificity seems to vary depending on the method used for scoring it, the methodology must always be taken into account when evaluating the symbiont range of a *Bdellovibrio* strain. As an illustration, the apparent symbiont range has been reported to be broader when it is determined by turbidity changes in liquid two-membered cultures than when it is scored by the ability to form plaques on lawns; moreover, when the cells both of the *Bdellovibrio* and of the substrate organism are washed and suspended in buffer, the symbiont range is said to be broader than with unwashed cells (Shilo and Bruff, 1965).

The following procedures have been used, or might be adapted to use, in studies on the symbiont range of bdellovibrios (the references are not meant to be exhaustive of the literature). (1) Phase contrast microscopic observations of attachments and other actions in wet mounts of bdello-vibrios and prospective associant bacteria (Stolp and Petzold, 1962; Stolp and Starr, 1963). (2) Formation by bdellovibrios of plaques or confluent lysis in lawns of associant bacteria on double-layer plates of dilute yeast extract peptone agar (Stolp and Petzold, 1962; Stolp and Starr, 1963). (3) Reduction in turbidity of two-membered broth cultures or of washed suspensions of both bacteria in non-nutrient buffer (Stolp and Starr, 1963; Shilo and Bruff, 1965). (4) Relative plaquing efficiency of *Bdellovibrio* strains on lawns of various substrate bacteria (Burger *et al.*, 1968; Klein and Casida, 1967; Taylor *et al.*, 1974). (5) An agar-block transfer procedure (Gillis and Nakamura, 1970). (6) The differential filtration procedure that has been used to score attachment kinetics (Varon and Shilo, 1968) might be adapted. Some of these procedures are further considered, in connexion with enumeration of *Bdellovibrio*, in Section V.

C. Possible ecological effects

In its natural habitats, *Bdellovibrio* quite likely functions as an antagonist of bacterial populations, both indigenous and exotic. Since these antagonized bacteria include some that are pathogenic to man and to plants and others that are important in soil processes of agronomic significance, there has been a great deal of interest in this subject (Dias and Bhat, 1965; Guélin *et al.*, 1968, 1969c; Keya and Alexander, 1975; Klein and Casida, 1967; Mitchell and Morris, 1969; Staples and Fry, 1973).

The possible use of bdellovibrios as "biological control" agents in infectious disease had early been broached (Stolp and Petzold, 1962; Stolp and Starr, 1963). Although some direct attempts along these lines have been made in animals (Nakamura, 1972) and in plants (Scherff, 1973), no results of practical utility have yet been achieved. This situation is reminiscent of the early work with bacteriophages; bacteriophages also have been tested— without much success of any lasting clinical significance—as potential therapeutic agents in infectious disease.

IV. ISOLATION OF SYMBIOSIS-COMPETENT *BDELLOVIBRIO* CLONES

Isolation of symbiosis-competent ("host-dependent" or "parasitic" or "predatory") bdellovibrios necessarily depends on the presence of susceptible associant ("host" or "prey") bacteria. However, several problems arise which stem from interference by or confusion with bacteriophages and other bacteriolytic agents, the relatively low population densities of *Bdellovibrio* in most natural habitats, and the occurrence in the same habitats of many other kinds of bacteria.

The lytic reactions caused by symbiosis-competent *Bdellovibrio* strains are similar in their overall operational manifestations to the lytic reactions caused by bacteriophage. Moreover, bdellovibrios and bacteriophages occupy the same ecological niche, the bacterial cell. Plating a mixture of bdellovibrios and (an excess of) associant bacteria on suitable agar media results in the formation of individual plaques or confluent lysis, just as in titration experiments with bacteriophages. In contrast to phage plaques, which are fully formed in less than one day, the lytic zones caused by *Bdellovibrio* increase in size over several days.

The concentration of bdellovibrios in natural habitats is relatively low as compared to many other bacteria. Moreover, by the very nature of their symbiotic mode of life, the known kinds of *Bdellovibrio* always occur in natural habitats together with other bacteria. Direct plating from natural materials onto lawns of prospective associant bacteria, therefore, usually is associated with overgrowth by the accompanying ("contaminant") micro-organisms some of which may be unsusceptible to *Bdellovibrio*; such over-growth may interfere with development of individual plaques on the lawn. For this reason, the initial steps in the isolation procedure must involve the prior separation of the bdellovibrios from the other microbes occurring in the same habitat. At the same time, interference by bacterio-phages or other bacteriolytic agents must be eliminated or minimized or at least recognized for what it is (see Section IV.E).

A. Isolation involving centrifugation and filtration

A relative enrichment of the bdellovibrios occurring in a suspension of natural materials may be accomplished by centrifugation and filtration. The rationale behind this procedure is the relatively low mass and small size of the bdellovibrios as compared to other bacteria. This technique (Procedure A) was originally used by Stolp and Starr (1963) for the isolation of bdello-vibrios from soil and sewage.

Procedure A:

1. Suspend 500 g of soil in 500 ml of tap water.
2. Shake vigorously for 1 h.
3. Centrifuge for 5 min at about $2000 \times g$.
4. Submit the supernatant (containing many types of microbes) to a series of filtrations using membrane filters of different porosities, starting with $3 \cdot 0 \ \mu m$ pore-size diameter and continuing with $1 \cdot 2$, $0 \cdot 8$, $0 \cdot 65$ and $0 \cdot 45 \ \mu m$ filters.
5. Plate the last two or three filtrates with prospective associant bacteria, using the conventional double-layer ("overlay") technique (Adams, 1959; Billing, this Series, Vol. 3B). The double-layer procedure as applied to *Bdellovibrio* involves first pouring into Petri dishes the bottom-layer agar and allowing it to solidify (in 9 cm diameter Petri dishes, 15 to 20 ml of an agar medium adequate for the substrate bacterium being used; agar con-centration is $1 \cdot 2$ to $1 \cdot 5\%$). Then the top-layer is prepared by adding, to 5 ml of molten top-layer agar (usually $0 \cdot 6\%$ agar) held at $42°C$, $0 \cdot 5$ ml of the suspension of substrate bacteria (10^9 cells or more per ml) and $0 \cdot 5$ ml of the appropriately diluted *Bdellovibrio* suspension. After these additions, the preparation is quickly and thoroughly mixed and poured onto the hardened bottom-layer so that the top-layer is uniformly spread. One must not allow the bdellovibrios to be exposed to temperatures above $42°C$ or to allow the preparation to harden prematurely. Because various brands of agar have differing gelling properties, the agar concentration of the top-layer should be chosen so that sharply defined and easily visible plaques are formed. Although it may be banal to mention this fact, we must emphasize that a too rigid top-layer will result in tiny plaques which are difficult to count, and

that a too soft top-layer will result in diffused margins and merging of the plaques.

Development of *Bdellovibrio* plaques is favoured by use of media of low nutrient content; e.g. yeast extract-peptone agar (YP; Stolp, 1965) composed of 0·3% Difco yeast extract, 0·06% Difco peptone, 0·05M Tris-buffer; pH 7·5. Top-layer agar contains 0·6% Bacto agar, bottom-layer agar contains 1·2% Bacto agar; these media are sterilized for 15 min at 121°C. The associant bacteria are grown on YGC agar slants of the following composition: 1% Difco yeast extract, 2% glucose, 2% $CaCO_3$ (finely divided), 1·5% agar, distilled water q.s. Sterilization of this medium is at 121°C for 15 min, the sterilized molten agar is swirled to suspend the $CaCO_3$ uniformly just before it is allowed to solidify into slants. The suspension of substrate bacteria is prepared by suspending the bacteria from one slant in 5 ml YP/10 (or other appropriate diluent).

6. Incubate for 2 to 4 days at a temperature appropriate both for the associant bacteria and the bdellovibrios, and check the lawns periodically for the presence of plaques.

7. Examine material from the plaques in wet mounts using the phase-contrast microscope. The actively motile and very small (about $0·3 \times 1·0$ μm) *Bdellovibrio* swarmers, as well as the bdelloplasted associant cells, are so very distinctive that they can be recognized at a glance.

8. Purify the isolated *Bdellovibrio* clones by means of Procedure C (see Section IV.C).

Bacteriophage plaques may develop when using YP medium, as multiplication of the associant bacteria is possible to some extent. Phage plaques ordinarily appear within 24 h, whereas *Bdellovibrio* plaques require 2 days or longer to develop. Easy differentiation is achieved by marking those plaques which develop early and considering only the later-developing plaques as possibly formed by *Bdellovibrio*.

By centrifugation and step-wise filtration, a considerable fraction of the bdellovibrios known to be present in the habitat is lost. Therefore, this technique is not suitable for quantitative enumeration of the bdellovibrios present in natural materials. The two techniques devised by Varon and Shilo (1966, 1968, 1970), differential centrifugation on a linear Ficoll gradient and a single passage through a 1·2 μm pore-size membrane filter, might be applied to Procedure A for increasing the proportion of bdellovibrios recovered and possibly for enumerating them from natural habitats (see Section V.D).

B. Enrichment with non-proliferating associant bacteria

The development of a specific enrichment technique (Stolp, 1968) has greatly facilitated the isolation of *Bdellovibrio*. This method (Procedure B) is based on the use of non-proliferating bacteria as prospective associants (10^{10} cells/ml; 1 day old) in very dilute (0·03%) yeast extract solution or in a non-nutrient buffered mineral salt solution. Under such conditions, the bacterial cells are the only or major source of nutrients, thus allowing selective development of the symbiosis-competent bdellovibrios. At the same time, the various other micro-organisms ("contaminants") present in the inoculum do not multiply for lack of nutrients, and the development of bacteriophages is considerably reduced because they can propagate only in proliferating bacterial cells, which are relatively rare in this system.

Procedure B:

1. Suspend prospective associant bacteria (grown for 24 h on YGC agar; final concentration 10^{10} bacteria/ml) either in 0·03% Difco yeast extract solution, or in non-nutrient buffered mineral salt solution containing 1000 ml 50 mM Tris-buffer (pH 7·5), 0·2 g $CaCl_2.2H_2O$, and 0·2 g $MgSO_4.7H_2O$; both solutions are sterilized by autoclaving at 121°C for 15 min.

2. Transfer 100 ml portions into Erlenmeyer flasks (300 ml) and add small amounts of the natural material which is to be checked for the presence of *Bdellovibrio*; e.g. 100 mg of soil, 0·5 ml of sewage, or 1 ml of river water.

3. Incubate on a rotary shaker for 2 to 4 days (30°C).

4. Check macroscopically for lysis (reduction of optical density) and microscopically (phase-contrast) for presence of the highly motile, tiny, usually vibrioid bdellovibrios.

5. If step 4 is positive, centrifuge 5 min at 2000 × g, filter through a 0·45 μm membrane filter, plate a dilution series on lawns of prospective associant bacteria for production of single plaques. If step 4 is negative, either continue incubation, or transfer 1 ml of the mixed culture into a fresh suspension of the associant bacteria (i.e., make a second enrichment).

6. Plating and purification are done as in Procedure A and Procedure C.

So far, there is no method known for enriching specifically for any one of the three presently recognized terrestrial species of *Bdellovibrio* (Burnham and Robinson, 1974); however, marine bdellovibrios require specific enrichment (see Section IV.D).

C. Purification of *Bdellovibrio* cultures

Starting from the cultures yielded by Procedure A or Procedure B, it is easy to make two-membered "pure cultures" (clones derived from individual cells) of symbiosis-competent bdellovibrios together with their associant bacteria by the following procedure, which is comparable to the conventional serial dilutions used with colony-forming bacteria.

Procedure C:

1. Material from a plaque is suspended in about 2 ml of YP (or YP/10; i.e. YP diluted ten-fold) solution.

2. The suspension is filtered through a membrane filter (0·45 μm; filter syringe), diluted serially with YP, and the serial dilutions are prepared into lawns of a suitable associant bacterium for production of individual plaques. Plating is done by mixing 0·5 ml of each serial dilution of the *Bdellovibrio* suspension, 0·5 ml of associant bacterial culture (in excess; about 10^{10}/ml), and 5 ml of top-layer agar (see Section IV.B, 5).

D. Isolation of marine bdellovibrios

The occurrence of bdellovibrios in marine environments—estuarine, coastal, and offshore—has been reported in various contexts by several investigators (Daniel, 1969; Guélin and Cabioch, 1970; Mitchell and Morris, 1969; Mitchell *et al.*, 1967; Shilo, 1966; Taylor *et al.*, 1974). Some of these studies were addressed to the fate of terrestrial bdellovibrios and terrestrial associant bacteria when they meet the sea. Others have dealt with true marine bdellovibrios, those that live in the oceans; it is this latter kind of *Bdellovibrio* that we consider here.

Several methodological points are raised in these studies and we will focus our attention here on those factors which seem to us most important for the isolation and characterization of genuinely marine bdellovibrios. Highly relevant are the reports to the effect that—like other marine bacteria (Reichelt and Baumann, 1974)—the *Bdellovibrio* strains isolated from off-shore marine environments not only tolerate 3% NaCl (625 mM) but actually require sodium ion. Taylor *et al.* (1974) report that all 13 *Bdellovibrio* strains isolated by them from the Pacific Ocean off the coast of Oahu, Hawaii, had an obligatory requirement for at least 75–100 mM NaCl, with an optimum at 125–150 mM NaCl. This sodium requirement was not satisfied by potassium. Ca^{2+} and Mg^{2+} also were required, as in terrestrial strains (Huang and Starr, 1973b; Seidler and Starr, 1969b).

Another pertinent fact is that bacteria of marine origin are more adequate associants for bdellovibrios isolated from offshore marine environments than are bacteria of terrestrial origin (Taylor *et al.*, 1974). This obvious ecological point is overlooked in many studies on *Bdellovibrio*, and the underlying principle bears repetition here. Symbiosis-competent bdello-vibrios are dependent for their very existence on other bacteria. Given the specificity of these associations (see Section II.B), it is quite likely that the most suitable associant bacteria in a given situation would be those indigenous to the habitat under study. This principle—often ignored—is well illustrated in the findings of Taylor *et al.* (1974) and it is involved in the procedure (Procedure D) we are recommending here for the isolation of marine bdellovibrios. Procedure D is based on satisfaction of the requirement of marine bdellovibrios for Na$^+$, their preference for marine bacteria as associants, and the likelihood that they would be favoured by incubation at relatively low temperatures.

Procedure D:

1. As a general guide, Procedures A, B and C might be used. However, at least one-fourth of the distilled water in the recipes for the various media and solutions must be replaced by seawater or artificial seawater. Alternatively, the formulations specified by Taylor *et al.* (1974) could be followed. At every step in the operation, the medium or solution must contain about 125 mM Na$^+$ ($\equiv 0.6\%$ NaCl).

2. The associant bacteria should be chosen from among the aerobic, Gram-negative marine bacteria which are common in the habitat being studied. The paper by Baumann *et al.* (1972) provides an excellent guide on this point.

3. Because many marine bacteria are adapted to living at relatively low temperatures, some attention must be devoted to determining the optimum temperature for the interbacterial association under investigation.

4. As the concentration of the bdellovibrios in marine habitats usually is extremely low, the application of Procedure B for enrichment of *Bdellovibrio* is the most promising technique. A pre-concentration by centrifugation (30 min, 8000–10,000 × g, +5°C) of large quantities of sea water (1 litre or even more) may be necessary.

E. Entities with which *Bdellovibrio* might, at first glance, be confused

The isolation of symbiosis-competent ("parasitic") bdellovibrios from natural habitats in general is based either on the development of lytic zones (plaques) in bacterial lawns (Petri dishes), or on the reduction of optical density in bacterial suspensions (test tubes or flasks). The appearance of plaques or the reduction of optical density, however, do not necessarily prove the presence of *Bdellovibrio* in primary isolations because bacteriophages, myxobacteria, and amoebae and other protozoa can cause very similar lytic reactions (see Sections IV.A, B). This fact has to be recognized in enumerating *Bdellovibrio* in natural habitats (see Section V.D).

The most reliable check for the presence of *Bdellovibrio* is microscopic control using phase-contrast microscopy. If agar material from a lytic zone (preferably from the edge of a plaque) or a drop from a lysing bacterial culture are examined in wet mounts, the highly motile, usually vibrioid and very tiny bdellovibrios are easily recognized. Occasionally (particularly in primary isolations from marine environments), confusion may arise from the presence of small and motile vibrioid bacteria other than *Bdellovibrio*. In such situations, it is advisable to check for the presence or absence of the attachments and "spheroplasts" ("bdelloplasts") which are characteristic of bdellovibrios.

V. ENUMERATION OF SYMBIOSIS-COMPETENT BDELLOVIBRIOS

Basically, there are two general methods that are feasible for enumerating symbiosis-competent (S-C) bdellovibrios: total counts by direct microscopic observation and viable counts based on the ability of S-C bdellovibrios to form plaques on lawns of associant bacteria. Each procedure has its advantages and disadvantages (see the Chapters by Mallette and by Postgate in Vol. 1 of this Series).

A. Direct microscopic counts

It is difficult and tedious to count particles as small as *Bdellovibrio* cells under the phase-contrast microscope. The Petroff–Hausser counting chamber (Mallette, this Series, Vol. 1) is a useful aid. Incorporation of glycerol into the cell suspensions to a final concentration of 25% (v/v) reduces streaming and Brownian movement (Snellen *et al.*, 1976). The preservation of *Bdellovibrio* suspensions with formalin, a common practice when using Petroff–Hausser counting chambers, increases the likelihood of cell aggregation (which would make the counts less precise); it is not known whether the addition of phosphate to the formalin (Norris and Powell, 1961) would solve this problem as it has in other systems. To improve the counting precision, each preparation must be loaded into the counting chamber several times and a relatively large number of fields must be counted in each replicate loading.

Since suspensions of S-C bdellovibrios propagated with associant bacteria usually represent mixtures of both organisms—unless separated by selective centrifugation and filtration—microscopic counting is rendered difficult if the size and shape of the two associants do not allow easy differentiation. There seems not to have been any use made to date of stained quantitative smears for direct microscopic enumeration of bdellovibrios.

B. Plaque counts

Viable counts in terms of colonies are, of course, impossible with S-C bdellovibrios, since they do not form visible colonies under any conditions that are presently known. Eventually, it may become possible to make such colony counts by using plating media which meet the nutritional requirements by adding either materials extracted from cells of associant bacteria (Horowitz et al., 1974) or defined substances known to be essential for growth and multiplication. However, at the moment, the only feasible viable counting procedure is the plaque assay method. As with any viable count, the numbers of Bdellovibrio cells calculated from plaque-forming units (PFU) are usually lower than the numbers in the same preparation calculated from direct microscopic (Petroff–Hausser) counts. However, by careful attention to the procedural details, it has been possible to attain plaque counts as high as 85 to 95% of the corresponding direct counts (Snellen et al., 1976).

Based on what is presently known, the medium of choice for plaque counts in most combinations of Bdellovibrio strains and associant bacterial strains is a yeast extract, peptone, sodium acetate, and cysteine (YPSC) agar supplemented with Ca^{2+} and Mg^{2+} (Huang and Starr, 1973b).

Cation-supplemented YPSC agar

Bacto yeast extract	0·1%
Bacto peptone	0·1%
Sodium acetate trihydrate	0·5%
L-cysteine hydrochloride	0·005%
$CaCl_2$	0·002M
$MgSO_4$	0·003M
Distilled water	to volume

pH adjusted before autoclaving to about 7·6 with 1N NaOH; top-layer agar (usually) has 0·6% to 0·8% Bacto agar and bottom layer agar has 1·5% Bacto agar; autoclaved at 121°C for 15 min. In some cases, the use of less rich nutrient media or even heavy suspensions of associant bacteria in non-nutrient buffered mineral salt agar may give better results.

Procedure E:

The Bdellovibrio suspensions (usually a fresh lysate) that are to be enumerated by plaque counts are diluted with 0·025M Tris buffer (pH 7·5) containing 0·002M $CaCl_2$ and 0·003M $MgSO_4$. Appropriate dilutions of the suspensions are plated on double-layer YPSC plates in which the top-layer was previously seeded with (usually) 1·0 ml of a 24-h old culture of the associant bacterium (concentration ca. 5×10^9 cells/ml). It goes without saying that the strain of associant bacterium must be one which is fully susceptible to the Bdellovibrio strain being enumerated. Many factors affect the plaquing efficiency (summarized in Starr and Huang, 1972). Among the more important are the relative and absolute densities of both the Bdellovibrio and the associant bacteria, the physiological condition of the Bdellovibrio suspension (the swarmers lose viability rapidly, often within a few hours), the metabolic products of the associant bacteria (since these will differ in different culture media and with different bacteria, it may be necessary to use in some systems lawns prepared in other than YPSC agar), the water content and thickness of the semi-solid agar top-layer, and the incubation temperature. Each of these factors, and others, must be evaluated in a given system and the conditions optimized. The plaques are usually fully developed after about 48–72 h, and they can then be counted directly, preferably in an obliquely-illuminated colony counter. To make the plaques more readily visible, the lawns might each be flooded at this time with 5 ml of a 0.1% (w/v) solution of 2,3,5-triphenyltetrazolium chloride in PY broth (Seidler and Starr, 1969b) and incubated either aerobically or under N_2 for 1–3 h at 37°C. Depending on the substrate organism and other factors, either the lawn or the plaques (or their interfaces) become(s) coloured—thus enhancing plaque visibility.

By varying the strains of associant bacteria used in the lawns, and keeping the *Bdellovibrio* suspensions and all other factors constant, one has a basis for determining the relative plaquing efficiency on the different associant bacteria, as already noted in Section III.B in connexion with the topic of symbiont specificity.

C. Other procedures related to enumeration

1. *Use of the Coulter counter*

As already noted, *Bdellovibrio* swarmers are quite small, about one-fifth to one-tenth the volume of ordinary bacterial cells such as stationary-phase *Escherichia coli*. This very small size should make the use of the ordinary Coulter counter (Kubitschek, this Series, Vol. 1) unreliable for enumerating bdellovibrios, especially in the presence of particulate debris resulting from dissolution of the bacterial associant cells. Nevertheless, Shilo and Bruff (1965) have used a Coulter counter for enumerating *Bdellovibrio starrii* strain A3.12. Some of the recent, more sophisticated versions and adaptations of this instrument might be tested for utility in the *Bdellovibrio* system; a recent note by Patinkin (1975) in fact reports the successful use, in sizing (and enumerating) N–S bdellovibrios, of a Coulter counter modified in several respects. It would indeed be useful to be able to use the Coulter counter for making differential counts of *Bdellovibrio* swarmers, unattacked cells of the associant bacteria, and bdelloplasted associant cells.

2. *Instrumented, membrane filter, direct count method*

A membrane filter method has been developed for making direct counts of *Bdellovibrio* (Snellen *et al.*, 1976).

A known volume of a quantitatively diluted suspension of *Bdellovibrio* cells in Tris-buffer is filtered through a known area of a 100 nm pore-size Millipore brand membrane filter. A clarification solvent renders the filter transparent so that the *Bdellovibrio* cells on the filter can be photographed and later counted in a Quantimet Image Analyzing Computer. The number of *Bdellovibrio* cells in the undiluted suspension is calculated from the observed number of cells per unit area of the filter, the dilution factor, and the volume of diluted suspension passed through the filter. When there are about 3·5 cells per 100 μm^2 of filter surface, the bdellovibrios are distributed on the filter in a Poisson manner. The direct counts obtained by this membrane filter method correlate well with viable counts based on plaque assays and with direct counts obtained by the Petroff–Hausser procedure.

This instrumented membrane filter procedure has certain advantages and disadvantages (Snellen *et al.*, 1976). The filter immobilizes all cells in the same focal plane, making feasible a permanent photographic record of the primary data. Although the overall procedure takes more time than does the use of the Petroff–Hausser chamber, the initial steps (filtration and photography) can be done rather quickly and the counting carried out later on a batch of photographs. Moreover, the use of an automatic image analyser is a particular asset when large-scale counting is to be effected. The major drawback of the procedure is that—unless the image analyser can be programmed to discriminate better—all contaminating particulate matter must be removed from the *Bdellovibrio* suspension, which is not always feasible. Although the photographic recordings can be counted by eye, the procedure requires an electronic image analyser for its optimal use and such instruments are usually not available.

3. *Miscellaneous procedures*

A variety of other methods used (or adaptable to use) for enumerating *Bdellovibrio* is noted in Section III.B and in Section V.D.

D. Enumeration of bdellovibrios as they occur in natural habitats

Most of the procedures considered in Sections V.A, B and C were devised for enumeration of bdellovibrios in laboratory cultures (or for other purposes). For a number of reasons, they generally are unsuited for enumeration of bdellovibrios as they occur in natural habitats. Direct plating on growing lawns, without centrifugation or filtration, from natural habitats usually yields a welter of confusion from lytic zones caused by bacteriophages, other bacteriolytic bacteria such as myxobacteria, and bacteriotrophic protozoa, as well as from overgrowth of the nutrient lawn by many sorts of microbes. On the other hand, when centrifugation and/or filtration steps are included, only a variable and usually small fraction of the bdellovibrios present in the natural material will be recovered. Enrichment procedures, by their very nature, are usually inappropriate for estimating directly the numbers of bdellovibrios in the original material prior to enrichment. Any single kind of bacterium used in lawns as the prospective associant can not be expected to be susceptible to every kind of *Bdellovibrio* that might be present in the natural material being studied; there is no known "universal" substrate organism that is susceptible to all of the already known *Bdellovibrio* strains—let alone to those that are still unknown.

Despite these problems, there have been attempts to enumerate bdellovibrios in natural environments. Klein and Casida (1967) used a procedure based on dilution-to-extinction, "amplification" by enrichment, and most-probable-number statistics. They prepared decimal serial dilutions of soil suspensions and to each dilution added cells of associant bacteria. After three days, the suspensions were plated on lawns for detection of plaques. By this technique, they determined which dilutions were at the borderline between capability and incapability of giving rise to *Bdellovibrio* plaques; from such data, the numbers of bdellovibrios were estimated using most-probable-number tables. This procedure might be improved by using as substrate organism, both in the enrichment and in the lawns, bacteria indigenous to the habitat being studied (rather than *Escherichia coli*, the bacterium which was used in this study).

One example of a study in which bdellovibrios in sewage and river water were enumerated by direct plating on growing lawns is that by Staples and Fry (1973). There was considerable interference by plaques caused by amoebae and myxobacteria, and attempts to eliminate the protozoa selectively by means of antiprotozoal agents were unsuccessful. Despite these difficulties, these experiments did allow a rough estimate to be made of the numbers of bdellovibrios present in these natural environments.

Some of the problems associated with direct plating could be reduced if the lawns were prepared with non-nutrient mineral salt agar (see Procedure B, Section IV.B), with the non-nutrient top-layer agar so heavily dosed with the associant bacteria that it is perceptibly turbid. A procedure somewhat along these lines was used by Keya and Alexander (1975) in enumerating from soil bdellovibrios active against *Rhizobium*. Their non-nutrient medium contained actidione, which seemed to work effectively in eliminating overgrowth by the ubiquitous soil fungi.

The usual centrifugation and filtration steps (as in Procedure A; see Section IV.A) result in losses of the bdellovibrios, sometimes to the extent of two or three orders of magnitude. It might be possible to achieve the benefits expected from the usual (procedure A) centrifugation and filtration, without such losses and in fact with good quantitative precision, by applying two procedures that have been devised by Varon and Shilo (1966,

1968, 1970). In one procedure (Varon and Shilo, 1966, 1970), the free bdellovibrios are separated from the associant bacteria by means of differential centrifugation through a linear Ficoll gradient. This method reportedly allowed recovery of about 80% of the bdellovibrios in the upper portion of the gradient, while 70–90% of the associant bacteria tested in this system moved to the bottom of the gradient. The procedure has not yet been checked systematically for separating bdellovibrios quantitatively from most other (larger) microbes occurring in the same natural materials. If this separation should be feasible, accurate estimations of *Bdellovibrio* numbers might be made routinely by plaque counts or direct microscopic counts of material taken from the upper parts of such Ficoll gradients. A related procedure, using dextran and polyethylene-glycol, has been reported by Eremenko and Mardachev (1972). The second technique (Varon and Shilo, 1968, 1970) involves a single filtration of a two-membered culture through a 1·2 μm pore-size membrane filter. In the laboratory system used, there was almost complete recovery of the bdello-vibrios in the filtrate while 90–95% of the associant bacteria were retained on the filter. Here again, evaluation needs to be made of the applicability of this simple procedure to precise enumeration of bdellovibrios as they occur in natural environments.

VI. ISOLATION OF NONSYMBIOTIC *BDELLOVIBRIO* CLONES

Stolp and Petzold (1962) reported the isolation of nonsymbiotic ("saprophytic") bdellovibrios from populations of the symbiosis-competent ("parasitic") wild-type bdellovibrios by massive inoculation of filtered lysates into complex nutrient medium (i.e., on nutrient agar plates). The procedure requires effective separation of the bdellovibrios from the residual associant bacteria of a lysate in order to avoid overgrowth of the bdello-vibrios by the colonies of the associant bacteria. Further use of this method was made by Stolp and Starr (1963).

Seidler and Starr (1969a) originated a technique which has been consistently successful in the isolation of nonsymbiotic ("host-independent") derivatives. They were able to demonstrate that every symbiosis-competent *Bdellovibrio* strain examined by them had the potentiality to yield non-symbiotic derivatives (frequency: 1 in 10^6 to 10^7). The technique is based on the selective development of nonsymbiotic derivatives in the presence of an antibiotic that does not act against the bdellovibrios but does exclude over-growth by associant bacteria. From the nonsymbiotic clones, upon exposure to suitable associant bacteria, revertants to the symbiosis-competent condition can be selected.

Procedure F:

This selection procedure (Seidler and Starr, 1969a) includes the following steps:

1. Selection of a streptomycin-resistant (sm^r) clone from the usually streptomycin-sensitive (sm^s) symbiosis-competent ("H-D"; "parasitic") wild-type *Bdellovibrio* strain.

2. Propagation of the S-C sm^r *Bdellovibrio* on sm^s associant bacteria in YP or YP/10 solution until lysis is complete.

3. Inoculation of the lysate (bdellovibrios and residual associant bacteria) into a complex agar medium containing streptomycin. Under such conditions, there is a selective development of nonsymbiotic sm^r *Bdellovibrio* mutants into colonies, whereas the residual sm^s associant cells are killed by the antibiotic, thus avoiding interference between *Bdellovibrio* and associant cell development. These nonsymbiotic ("saprophytic"; "H-I") *Bdellovibrio* strains grow well in peptone yeast extract (PY) broth or agar media (Stolp and Starr, 1963). Composition of PY broth: 1000 ml distilled water, 10 g Difco peptone, 3 g Difco yeast extract; pH 7·2.

Isolation of such nonsymbiotic bdellovibrios directly from natural habitats has not yet been reported; they are known only by derivation from symbiosis-competent *Bdellovibrio* clones. The properties of N-S *Bdellovibrio* strains have been reported, in various contexts, by Seidler *et al.* (1969, 1972), Seidler and Starr (1969a), Stolp and Petzold (1962), Stolp and Starr (1963), and others. Additional facts about them can be gleaned from Section VIII.

VII. MAINTENANCE OF *BDELLOVIBRIO* CULTURES

Cultures of symbiosis-competent bdellovibrios may be maintained by regular transfer of the organism into suspensions or onto lawns of suitable associant bacteria. The rather rapid loss in viability of *Bdellovibrio* (caused by its very active endogenous respiration; Rittenberg and Shilo, 1970; Hespell *et al.*, 1973) requires transfers at fairly frequent intervals, as often as every week or two in particular systems.

Lyophilization of skim-milk suspensions of both S-C and N-S *Bdellovibrio* cells has been successfully practiced as a conservation procedure; the results, however, are sometimes erratic (Starr and Huang, 1972). There are as yet no reports on the survival of *Bdellovibrio* cultures frozen in liquid nitrogen.

Safe maintenance of bdellovibrios is possible in deep-frozen condition. The following procedure (Stolp, 1973) is recommended:

Procedure G:

1. Transfer 5 ml of a lysate from a liquid two-membered culture of

Bdellovibrio plus its associant (in YP/10) into a sterilized screw-cap test tube and add 1 ml of a fresh culture of associant bacteria. The lysate should contain at least 10^9 actively motile *Bdellovibrio* swarmers per ml; the associant bacteria (24 h old) taken from a slant and suspended in YP/10 solution should be added to give a final concentration of approximately 10^9 cells per ml.

2. Incubate the mixture of bdellovibrios and associant bacteria for 30 min on a shaker. At the end of this incubation period, many cells of the associant organism have been invaded by *Bdellovibrio* and these are located in the periplasmic spaces of the associant bacteria.

3. Add glycerol (final concentration: 10%, v/v) and store the preparations at $-30°C$.

4. To restore viability, prepare a double-layer plate containing in the top-layer suitable associant bacteria, melt a bit of the frozen *Bdellovibrio* preparation by applying a finger to the tube in which it is contained, transfer a drop or a loopful to the bacterial lawn, and incubate. Assure purity by applying Procedure C. The revival of the frozen specimen can also be achieved by inoculating some of it into a fresh associant culture suspended in YP/10 solution.

Cultures subjected to Procedure G have been recovered in good condition after 5 years of storage.

VIII. OTHER METHODS FOR STUDYING *BDELLOVIBRIO*

Albeit the bdellovibrios are unusual bacteria in certain respects, they are after all bacteria. Hence, the laboratory study of bdellovibrios—especially of N-S strains—has involved application and adaptation of methods commonly used by bacteriologists. There is no point in detailing such methods here (they generally can be traced through the review articles cited in Section I). Instead, it seems to us useful to summarize now, in brief vignettes, the methodological situation regarding *Bdellovibrio* in several conventional areas of bacteriology that have not already been treated in the previous Sections of this Chapter. This skeleton presentation—accompanied as it is by non-exhaustive citations to the historical and current primary literature—will also try to point out methodological and factual deficiencies in the hope that the community of *Bdellovibrio* workers will thereby be stimulated to find imaginative solutions.

A. Morphology, morphogenesis and ultrastructure

Considerable information has been acquired by means of microscopy about the morphology and ultrastructure of *Bdellovibrio*, its association with other bacteria, and its dimorphic life-cycle. Phase-contrast microscopy has been an indispensable tool from the start (Ross *et al.*, 1974; Scherff *et al.*, 1966; Seidler and Starr, 1969a; Shilo and Bruff, 1965; Starr and Baigent, 1966; Stolp and Petzold, 1962; Stolp and Starr, 1963). Coupled with high-speed cinematography (Stolp, 1967a, 1967b), phase-contrast microscopy has provided a great deal of information about the interbacterial association.

Electron microscopy—usually involving transmission microscopy, but more recently scanning microscopy (Snellen and Starr, 1976)—has been carried out with whole cells (Abram and Davis, 1970; Stolp and Petzold, 1962), flagellar preparations (Seidler and Starr, 1968), thin sections (Abram *et al.*, 1974; Burger *et al.*, 1968; Burnham *et al.*, 1970; Starr and Baigent, 1966; Stolp, 1968), and freeze-fractured and freeze-etched preparations (Abram and Davis, 1970; Snellen and Starr, 1974, 1976).

Much of the work on morphogenesis has been descriptive. Recently, a modified Coulter counter has been used to size N-S *Bdellovibrio* cells, and Virginiamycin S has been used to secure uniform populations of the helical forms (Patinkin, 1975).

B. Metabolism, growth, nutrition and genetics

Enzymological and biochemical procedures used for *Bdellovibrio* research have been quite standard (Engelking and Seidler, 1974; Fackrell and Robinson, 1973; Gloor *et al.*, 1974; Huang and Starr, 1973a; Simpson and Robinson, 1968; Steiner *et al.*, 1973). Studies on energy metabolism also have involved mainly conventional procedures (Gadkari and Stolp, 1975, 1976; Hespell *et al.*, 1973; Rittenberg and Hespell, 1975).

Growth studies of S-C bdellovibrios, including single-step growth curves and burst-size, have employed well-known techniques (Pritchard *et al.*, 1975; Rittenberg and Langley, 1975; Seidler and Starr, 1969b; Varon and Shilo, 1968, 1969). No S-C or N-S *Bdellovibrio* strain has yet been cultivated in a fully defined medium, although there are reports on the fairly complex preparations extracted from other microbes which are essential for the growth of S-C or N-S strains (Horowitz *et al.*, 1974; Ishiguro, 1973, 1974; Reiner and Shilo, 1969).

Molecular genetic procedures (GC contents and nucleic acid hybridization *in vitro*) have been applied to both N-S and S-C bdellovibrios, with the only unusual feature being the relatively minor difficulty of securing clean nucleic acid preparations of S-C bdellovibrios free from contaminating substances originating from the substrate bacteria (Seidler *et al.*, 1969, 1972).

Mutant strains of *Bdellovibrio* have been isolated and used for several purposes. The reversible shift between S-C and N-S forms is usually assumed to be a mutational event (Seidler and Starr, 1969a; Shilo, 1973b; Stolp and Petzold, 1962; Stolp and Starr, 1963; Varon *et al.*, 1974).

Streptomycin-resistant mutant clones of S-C bdellovibrios are a valuable aid in selecting N-S forms (Seidler and Starr, 1969a). Temperature-sensitive "attachment" and "penetration" mutants of S-C *Bdellovibrio* strains have been described (Dunn *et al.*, 1974). None of these systems has been analysed by the usual techniques of bacterial genetics, nor are recombinational genetic studies of *Bdellovibrio* now possible because neither the S-C nor the N-S forms can yet be cultivated in a fully defined medium. This lack makes it quite impossible to use the recombinational genetic techniques ordinarily employed (assuming, of course, that workable conjugational, transductional, and/or transformational systems can be defined in *Bdellovibrio*) and no one has yet come up with alternative procedures.

One aspect of the behavioural physiology of *Bdellovibrio*, understanding of which may possibly be limited by methodological deficiencies, is the seeming "recognition" by *Bdellovibrio* swarmers of congenial symbionts and the rapid locomotion of the swarmers directly toward them. One might reasonably expect that such "recognition" (if it exists) and the locomotion would be mediated by a chemotaxis, and that use of the Adler (1966, 1973) capillary-tube technique with congenial symbionts as "bait" would provide useful information on this point. Unfortunately, the latter is not the case: summarizing unpublished observations made in his laboratory and elsewhere, Starr (1975b) remarks that although there are sometimes "statistically valid increases in numbers of bdellovibrios moving toward congenial bacteria, used as bait [in the Adler procedure], generally rather small numbers of bdellovibrios swam even to the congenial bacterial bait; moreover, the results in replicate experiments have sometimes been contradictory". The only publication on this subject (Straley and Conti, 1974) comes to much the same conclusion, but it does report that bdellovibrios exhibit (in the Adler technique) an unmistakable positive chemotaxis to some components of yeast extract. In the present state of ignorance, it is by no means clear whether the Adler technique *per se* is for some unknown reason inadequate as it has been applied to this interbacterial system or whether there is an anthropomorphic conceptual and/or perceptual error inherent in the notion that there is "recognition" by *Bdellovibrio* of congenial substrate bacteria and a chemotactically-mediated locomotion toward them.

C. Bacteriophages

Like other bacteria, the bdellovibrios are susceptible to bacteriophages—so-called "bdellophages". The existence of N-S *Bdellovibrio* strains has greatly facilitated bdellophage isolation. The first reported *Bdellovibrio* phage was isolated by Hashimoto *et al.* (1970). Several others were isolated and studied by Althauser *et al.* (1972), by Schindler and Ludwik (1972), and by Varon and Levisohn (1972). Varon and Levisohn (1972) and Varon (1974) studied a three-membered system consisting of a bdellophage (active against *Bdellovibrio*), a strain of S-C *Bdellovibrio* (sensitive to the bdellophage and active against *Escherichia coli*), and *E. coli* itself. Whereas, in an ordinary phage system, the action of the virulent phage results in the development of clear zones of lysis (plaques) in a turbid bacterial lawn, the three-membered system is characterized by growth of bacterial colonies in the translucent (confluently lysed) lawn.

All of the presently known bdellophages are virulent and they have originally been isolated using N-S *Bdellovibrio* strains. Some of these, as already noted, are active against the parental S-C bdellovibrios in a three-membered system. In order to demonstrate bdellophage action in a three-membered system, it is necessary to pay attention to the following experimental conditions: (a) the concentration of the bdellovibrios and their substrate organisms must be adjusted in such a way as would result in confluent lysis; and (b) the concentration of the bdellophage must be adjusted so that there would be individual, well-separated phage particles deposited in the top-layer (which also contains the substrate bacteria and the S-C bdellovibrios). Under these conditions, the outward manifestations of phage infection are "negative plaques"; that is, the growth of colonies of the substrate organism that is protected from *Bdellovibrio* lysis by the local destruction by the bdellophage of the bdellovibrios.

D. Taxonomy

There are two major unsettled taxonomic issues concerning the bdellovibrios: the relationships of the genus *Bdellovibrio* to other groups of bacteria, and the intrageneric arrangement of the *Bdellovibrio* strains. The first issue, treated by several writers (Krieg and Hylemon, 1976; Starr and Seidler, 1971; Stolp, 1973; Stolp and Starr, 1963), results in tentative placement of *Bdellovibrio* in the taxonomic vicinity either of the genus *Spirillum* or the genus *Vibrio*—with the preponderance of evidence favouring the former affinity. The second issue, the delineation of species within the genus *Bdellovibrio*, has involved to date mainly molecular genetic and comparative enzymological techniques on the basis of which three species have been recognized: *Bdellovibrio bacteriovorus*, *Bdellovibrio stolpii*, and *Bdellovibrio starrii* (Burnham and Robinson, 1974; Seidler *et al.*, 1972). Some of the marine bdellovibrios (Taylor *et al.*, 1974) may well be members of still unnamed taxa different from the three named *Bdellovibrio* species. It would certainly be useful to conduct a full-spectrum taxonomic study of these organisms. The ease with which N-S clones can be selected and studied by a variety of procedures (many of which are not applicable to S-C

242   MORTIMER P. STARR AND HEINZ STOLP

clones in two-membered systems) suggests that it would be productive to use N-S clones for much of this systematic work.

As is the case in other bacterial groups, clarification of the taxonomic position—generic as well as specific—of the bdellovibrios may be made by comparative phage susceptibility studies. The same is true for comparative serology, a technique which in other groups of bacteria has been a powerful systematic tool, but which remains unexploited in *Bdellovibrio* research.

REFERENCES

Abram, D., and Davis, B. K. (1970). *J. Bact.*, **104**, 948–965.
Abram, D., Castro e Melo, J., and Chou, D. (1974). *J. Bact.*, **118**, 663–680.
Adams, M. H. (1959). "Bacteriophages". Interscience, New York.
Adler, J. (1966). *Science*, **153**, 708–716.
Adler, J. (1973). *J. Gen. Microbiol.*, **74**, 77–91.
Althauser, M., Samsonoff, W. A., Anderson, C., and Conti, S. F. (1972). *J. Virol.*, **10**, 516–524.
Baumann, L., Baumann, P., Mandel, M., and Allen, R. D. (1972). *J. Bact.*, **110**, 402–429.
Buckmire, F. L. A. (1971). *Bact. Proc.*, p. 43.
Buckmire, F. L. A., and Murray, R. G. E. (1970). *Can. J. Microbiol.*, **16**, 1011–1022.
Buckmire, F. L. A., and Murray, R. G. E. (1973). *Can. J. Microbiol.*, **19**, 59–66
Burger, A., Drews, G., and Ladwig, R. (1968). *Arch. Mikrobiol.*, **61**, 261–279.
Burnham, J. C., and Robinson, J. (1974). *In* "Bergey's Manual of Determinative Bacteriology", Eighth edition, pp. 212–214 (Eds. R. E. Buchanan and N. E. Gibbons), Williams & Wilkins Company, Baltimore.
Burnham, J. C., Hashimoto, T., and Conti, S. F. (1970). *J. Bact.*, **101**, 997–1004.
Crothers, S. F., Fackrell, H. B., Huang, J. C.-C., and Robinson, J. (1972). *Can. J. Microbiol.*, **18**, 1941–1948.
Daniel, S. (1969). *Rev. int. Océanogr. Méd.*, **15–16**, 61–102.
de Bary, A. (1879). "Die Erscheinung der Symbiose", Verlag von Karl J. Trübner, Strassburg.
Dias, F. F., and Bhat, J. V. (1965). *Applied Microbiol.*, **13**, 257–261.
Diedrich, D. L., Denny, C. F., Hashimoto, T., and Conti, S. F. (1970). *J. Bact.*, **101**, 989–996.
Dunn, J. E., Windom, G. E., Hansen, K. L., and Seidler, R. J. (1974). *J. Bact.*, **117**, 1341–1349.
Engelking, H. M., and Seidler, R. J. (1974). *Arch. Microbiol.*, **95**, 293–304.
Eremenko, V. V., and Mardachev, S. R. (1972). *C. r. Acad. Sci. D*, **274**, 1589–1592.
Fackrell, H. B., and Robinson, J. (1973). *Can. J. Microbiol.*, **19**, 659–666.
Finance, C., Schwartzbrod, L., and Baldo, S. (1974). *Revue de l'Institut Pasteur de Lyon*, **7**, 207–231.
Gadkari, D., and Stolp, H. (1975). *Arch. Microbiol.*, **102**, 179–185.
Gadkari, D., and Stolp, H. (1976). *Arch. Microbiol.*, **108**, 125–132.
Gillis, J. R., and Nakamura, M. (1970). *Infection and Immunity*, **2**, 340–341.
Gloor, L., Klubek, B., and Seidler, R. J. (1974). *Arch. Microbiol.*, **95**, 45–56.
Gromov, B. V., and Mamkaeva, K. A. (1972). *Tsitologiia*, **14**, 256–260.
Guélin, A., and Cabioch, L. (1970). *C. r. Acad. Sci. D*, **271**, 137–140.
Guélin, A., and Lamblin, D. (1966). *Bull. Acad. Nat. Méd. (Paris)*, **150**, 526–532.

Guélin, A., Lépine, P., and Lamblin, D. (1967). *Ann. Inst. Pasteur*, **113**, 660–665.
Guélin, A., Lépine, P., and Lamblin, D. (1968). *Rev. int. Océanogr. Méd.*, **10**, 221–227.
Guélin, A., Lépine, P., and Lamblin, D. (1969a). *Int. Vereinig. theor. angew. Limnol. Verh.*, **17**, 744–746.
Guélin, A., Lépine, P., and Lamblin, D. (1969b). *C. r. Acad. Sci., Paris*, **268**, 2828–2830.
Guélin, A., Lépine, P., Lamblin, D., and Petitprez, A. (1969c). *Bulletin Francais de Pisciculture*, **233**, 101–107.
Hashimoto, T., Diedrich, D. L., and Conti, S. F. (1970). *J. Virol.*, **5**, 97–98.
Hespell, R. B., Rosson, R. A., Thomashow, M. F., and Rittenberg, S. C. (1973). *J. Bact.*, **113**, 1280–1288.
Horowitz, A. T., Kessel, M., and Shilo, M. (1974). *J. Bact.*, **117**, 270–282.
Huang, J. C.-C., and Starr, M. P. (1973a). *Arch. Mikrobiol.*, **89**, 147–167.
Huang, J. C.-C., and Starr, M. P. (1973b). *Antonie van Leeuwenhoek J. Microbiol. Serol.*, **39**, 151–167.
Ishiguro, E. E. (1973). *J. Bact.*, **115**, 243–252.
Ishiguro, E. E. (1974). *Can. J. Microbiol.*, **20**, 263–265.
Keya, S. O., and Alexander, M. (1975). *Soil Biol. Biochem.*, **7**, 231–237.
Klein, D. A., and Casida, L. E., jr. (1967). *Can. J. Microbiol.*, **13**, 1235–1241.
Krieg, N. R., and Hylemon, P. B. (1976). *Ann. Rev. Microbiol.*, **30**, 303–325.
Maugeri, T. L. (1969). *Boll. Ist. Sieroterap. Milan*, **48**, 84–95.
Mishustin, E. N., and Nikitina, E. S. (1970). *Bull. Acad. Sci. U.S.S.R., Biol. Ser.*, **3**, 423–426.
Mitchell, R., and Morris, J. C. (1969). *In* "Proceedings of the 4th International Conference on Advances in Water Pollution Research", Prague, pp. 936. Pergamon Press, Oxford and New York.
Mitchell, R., Yanofsky, S., and Jannasch, H. W. (1967). *Nature*, **215**, 891–893.
Nakamura, M. (1972). *Am. J. Clin. Nutrition*, **25**, 1441–1451.
Norris, K. P., and Powell, E. O. (1961). *J. roy. Microsc. Soc.*, **80**, 107–119.
Ottow, J. C. G. (1969). *Landbouwkundig Tijdschrift*, **81**, 275–280.
Parker, C. A., and Grove, P. L. (1970). *J. Appl. Bact.*, **33**, 253–255.
Patinkin, D. (1975). *J. Bact.*, **124**, 564–566.
Pritchard, M. A., Langley, D., and Rittenberg, S. C. (1975). *J. Bact.*, **121**, 1131–1136.
Reichelt, J. L., and Baumann, P. (1974). *Arch. Microbiol.*, **97**, 329–345.
Reiner, A. M., and Shilo, M. (1969). *J. Gen. Microbiol.*, **59**, 401–410.
Rittenberg, S. C., and Hespell, R. B. (1975). *J. Bact.*, **121**, 1158–1165.
Rittenberg, S. C., and Langley, D. (1975). *J. Bact.*, **121**, 1137–1144.
Rittenberg, S. C., and Shilo, M. (1970). *J. Bact.*, **102**, 149–160.
Ross, E. J., Robinow, C. F., and Robinson, J. (1974). *Can. J. Microbiol.*, **20**, 847–851.
Scherff, R. H. (1973). *Phytopathol.*, **63**, 400–402.
Scherff, R. H., DeVay, J. E., and Carroll, T. W. (1966). *Phytopathol.*, **56**, 627–632.
Schindler, J., and Ludwik, J. (1972). *Acta virol. Tchecosl.*, **16**, 501–502.
Seidler, R. J., and Starr, M. P. (1968). *J. Bact.*, **95**, 1952–1955.
Seidler, R. J., and Starr, M. P. (1969a). *J. Bact.*, **97**, 912–923.
Seidler, R. J., and Starr, M. P. (1969b). *J. Bact.*, **100**, 769–785.
Seidler, R. J., Starr, M. P., and Mandel, M. (1969). *J. Bact.*, **100**, 786–790.
Seidler, R. J., Mandel, M., and Baptist, J. M. (1972). *J. Bact.*, **109**, 209–217.

Shilo, M. (1966). *Sci. J.*, **2**, 33–37.
Shilo, M. (1969). *Current Topics Microbiol. Immunol.*, **50**, 174–204.
Shilo, M. (1973a). *In* "Dynamic Aspects of Host-Parasite Relationships", Vol. 1, pp. 1–12, (Eds. A. Zuckerman and D. W. Weiss). Academic Press, New York and London.
Shilo, M. (1973b). *Bull. Inst. Pasteur*, **71**, 21–31.
Shilo, M., and Bruff, B. (1965). *J. Gen. Microbiol.*, **40**, 317–328.
Simpson, F. J., and Robinson, J. (1968). *Can. J. Biochem.*, **46**, 865–873.
Snellen, J. E., and Starr, M. P. (1974). *Arch. Microbiol.*, **100**, 179–195.
Snellen, J. E., and Starr, M. P. (1976). *Arch. Microbiol.*, **108**, 55–64.
Snellen, J. E., Marr, A. G., and Starr, M. P. (1976). *J. Appl. Bact.* In Press.
Staples, D. G., and Fry, J. G. (1973). *J. Appl. Bact.*, **36**, 1–11.
Starr, M. P. (1975a). *In* "Symbiosis", Symposia, Society for Experimental Biology, Vol. 29, pp. 1–20 (Eds. D. J. Jennings and D. L. Lee). Cambridge University Press, Cambridge.
Starr, M. P. (1975b). *In* "Symbiosis", Symposia, Society for Experimental Biology, Vol. 29, pp. 93–124 (Eds. D. J. Jennings and D. L. Lee). Cambridge University Press, Cambridge.
Starr, M. P., and Baigent, N. L. (1966). *J. Bact.*, **91**, 2006–2017.
Starr, M. P., and Huang, J. C.-C. (1972). *Advan. Microbial Physiol.*, **8**, 215–261.
Starr, M. P., and Seidler, R. J. (1971). *Ann. Rev. Microbiol.*, **25**, 649–678.
Steiner, S., Conti, S. F., and Lester, R. L. (1973). *J. Bact.*, **116**, 1199–1211.
Stolp, H. (1965). *Zbl. Bakt. Parasitenkunde, I. Abt., Suppl. Heft*, **1**, 52–56.
Stolp, H. (1967a). Film E-1314. Inst. für den wiss. Film, Göttingen.
Stolp, H. (1967b). Film C-972. Inst. für den wiss. Film, Göttingen.
Stolp, H. (1968). *Naturwissenschaften*, **55**, 57–63.
Stolp, H. (1973). *Ann. Rev. Phytopathol.*, **11**, 53–76.
Stolp, H., and Petzold, H. (1962). *Phytopathol. Z.*, **45**, 364–390.
Stolp, H., and Starr, M. P. (1963). *Antonie van Leeuwenhoek J. Microbiol. Serol.*, **29**, 217–248.
Stolp, H., and Starr, M. P. (1965). *Ann. Rev. Microbiol.*, **19**, 79–104.
Straley, S. C., and Conti, S. F. (1974). *J. Bact.*, **120**, 549–551.
Taylor, V. I., Baumann, P., Reichelt, J. L., and Allen, R. D. (1974). *Arch. Microbiol.*, **98**, 101–114.
Uematsu, T., Shiomi T., and Wakimoto, S. (1971). *Ann. Phytopathol. Soc. Japan*, **37**, 52–57.
Varon, M. (1974). *CRC Crit. Rev. Microbiol.*, **4**, 221–241.
Varon, M., and Levisohn, R. (1972). *J. Virol.*, **9**, 519–525.
Varon, M., and Shilo, M. (1966). *Israel J. Med. Sci.*, **2**, 654.
Varon, M., and Shilo, M. (1968). *J. Bact.*, **95**, 744–753.
Varon, M., and Shilo, M. (1969). *J. Bact.*, **97**, 977–979.
Varon, M., and Shilo, M. (1970). *Rev. int. Océanogr. Méd.*, **18–19**, 145–152.
Varon, M., Dickbuch, S., and Shilo, M. (1974). *J. Bact.*, **119**, 635–637.

Subject Index

A

Achromobacter, 76
Adenovirus Type 5, 191
Aeromonas proteolytica, 6, 11
Agar-strip machine, 18
Amines
    produced by micro-organisms, 79
    gas chromatographic separation of, 81
Aminopeptidases,
    activity in fungi, 4
    assay, preparation of yeast cell sus-
        pension for, 8
    cell-free, from *Enterobacteriaceae*, 10
    differentiation by assay of microbial, 1
    preparation of amino-acid β-naph-
        thylamide substrates, 9
    profile determination with, 10
    profile determination with cell-free,
        11
    profile of four yeasts, 3
    substrate specificities of,
        measured in whole cells, 8
        measured with soluble enzymes, 10
Ammonium molybdate, 187, 189, 191
Amoebae and *Bdellovibrio*, 230
Anisakis spp. 2
API, 36
Apiezon L (APL), 58
Apohaemocyanins, 192
Arthrobacter spp., 6
Ascaris suum, 2
Aspergillus flavus, 2, 4
A. niger, 4
A. oryzae, 4
A. parasiticus, 4
Associations, organismic, 219
Autoline system, 15
    contrast enhancement by oblique angle
        light sensing position, 34
    diffusion barriers for, 25
    diffusion centres for, 22
    growth factors, antibiotics and other
        chemicals used in, 39
    linear scanning for, 18
    measuring system for, 29

optical study of microcolony develop-
    ment, 31
scattered light measurements for
    growth determination in, 28
scattered light responses, 31
segmenting machine for, 26
signal analysis, 35
Auxotyping, of gonococci, 37

B

Bacillus, 64
B. anthracis, 6
B. brevis, 6
B. cereus, 6
B. circulans, 6
B. globigii, 6
B. laterosporus, 6
B. lentus, 6
B. licheniformis, 6
B. megaterium, 6
B. mesentericus, 6
B. polymyxa, 6
B. pumilus, 6
B. rotans, 6
B. sphaericos, 6
B. subtilis, 6
Bacterial pili—antibodies against, 195
Bacteroides, 63
B. amylophilus, 6
Bacteriophage PM2, 193
Bacteriophages and *Bdellovibrio*, 230
Basic fermentation end products, analy-
    sis of, 77, 78
Bdellophage, 240
Bdellovibrio, 218 *et seq.*
    bacteriophage of, 240
B. bacteriovorus, 241
B. chlorellavorus, 223
    clones, isolation of symbiosis com-
        petent, 225
    ecological effects of, 225
    enumeration of, 234
    methodology, 217 *et seq.*
    plaques, 235

DATE DUE